AF353007

Productive and Ecological Aspects of Mixed Cropping System

Editors

Anna Wenda-Piesik
Agnieszka Synowiec

MDPI • Basel • Beijing • Wuhan • Barcelona • Belgrade • Manchester • Tokyo • Cluj • Tianjin

Editors

Anna Wenda-Piesik
Department of Agronomics,
Faculty of Agriculture and
Biotechnology, UTP
University of Science
and Technology
Poland

Agnieszka Synowiec
Department of Agroecology
and Crop Production,
University of Agriculture in Kraków
Poland

Editorial Office
MDPI
St. Alban-Anlage 66
4052 Basel, Switzerland

This is a reprint of articles from the Special Issue published online in the open access journal *Agriculture* (ISSN 2077-0472) (available at: https://www.mdpi.com/journal/agriculture/special_issues/Mixed_Cropping).

For citation purposes, cite each article independently as indicated on the article page online and as indicated below:

LastName, A.A.; LastName, B.B.; LastName, C.C. Article Title. *Journal Name* **Year**, *Volume Number*, Page Range.

ISBN 978-3-0365-3895-2 (Hbk)
ISBN 978-3-0365-3896-9 (PDF)

Cover image courtesy of Katarzyna Pużyńska.

Contents

About the Editors

Anna Wenda-Piesik Professor, is an academic scientist employed at the Faculty of Agriculture and Biotechnology, Bydgoszcz University of Science and Technology. She is a specialist in sustainable agriculture and horticulture. Her particular scientific interests are biodiversity in agroecology; the integrated field production of wheat, rape, maize, and soybeans; and pest management in agricultural systems. The author has spent over 10 years researching the ecological and production attributes of cereal and cereal-legume mixtures, contributing to numerous publications and presentations. In the last two years, she has made several significant scientific achievements, especially in the production of soybean in Poland, which she represents as a research and teaching associate in the European Innovation Partnership and as the member of the Polska Soybean Society. She has managed four research projects and has provided numerous expert opinions and implementations for sustainable agriculture and food production practices. The effects of her completed projects have been disseminated in publications with a total IF of 68.6 and H index of 14. She is a member of committees for five European scientific journals and an expert in four scientific competition agencies.

Agnieszka Synowiec PhD, is an academic teacher at the Faculty of Agriculture and Economics. Her teaching and scientific interests focus on agroecology, weed science, and allelopathy. Specifically, she is interested in studying interspecies competition in crop mixtures and between crops and weeds.

 *agriculture*

MDPI

Editorial

Productive and Ecological Aspects of Mixed Cropping System

Anna Wenda-Piesik [1,*] and Agnieszka Synowiec [2]

[1] Department of Agronomics, Faculty of Agriculture and Biotechnology, UTP University of Science and Technology, Al. Kaliskiego 7, 85-796 Bydgoszcz, Poland

[2] Department of Agroecology and Crop Production, The University of Agriculture in Krakow, Al. Mickie wicza 21, 31-120 Krakow, Poland; a.synowiec@urk.edu.pl

* Correspondence: apiesik@utp.edu.pl

Citation: Wenda-Piesik, A.; Synowiec, A. Productive and Ecological Aspects of Mixed Cropping System. *Agriculture* **2021**, *11*, 395. https://doi.org/10.3390/agriculture11050395

Received: 9 April 2021
Accepted: 23 April 2021
Published: 27 April 2021

Publisher's Note: MDPI stays neutral with regard to jurisdictional claims in published maps and institutional affiliations.

Mixed cropping, also known as inter-cropping, polyculture, or co-cultivation, is a type of plant production system that involves planting two or more species (or cultivars) simultaneously in the same field in a variable order (row or rowless). Mixed cropping plays an important role in sustainable agriculture by adding value to crop rotations and agroecosystems. Various species provide complimentary use of environmental resources in a mixture in contrast to pure stands. The recent findings confirmed the benefits of intercropping, such as increased nitrogen uptake by cereals cultivated in a mixture with legumes [1–3], and more efficient use of water and nutrients in soil profile [4]. Different species in the mixtures also use field space more effectively [5], i.e., in weather conditions unfavorable for the growth of one species, the companion crop usually grows better [6,7]. Mixed systems are characterized by higher yields than pure stands [3,5,7,8]. Crop mixtures provide several agroecosystem services, e.g., increase biodiversity and support the diversity of beneficial insects, as well as reduce the outbreak of pests, which, among other things, is linked to decreased availability of food sources [9].

Mixed cultivation fully supports the various arguments presented in its favor. The latest research shows that, in the cropping of maize with common beans or garden nasturtium *Tropaeolum majus*, yields of dry matter were obtained in comparable quantities and qualities to those resulting from the cultivation of maize alone. This study showed that the intercropping of maize in Central Europe with flowering partners could be a suitable alternative to growing maize alone and can increase field biodiversity [7].

Scientific investigations on environmentally friendly mixed cropping should be supported by studies on the direct costs and long-term benefits that are most relevant to farmers. Mixed cropping plays an important role in sustainable and organic agriculture by increasing biodiversity in the crop rotation and agroecosystem, particularly in organic farming in mountainous areas. Spring cereals' intercropping increases the land equivalent ratio (LER) compared to the integrated farming [5]. The profits of grain yield in spring cereal mixtures with barley exceeded the other spring cereal species, such as oats and triticale, in their mixed and mono-crop cultures [5].

While utilizing polyculture systems, the farmers can keep their fields under continuous production and enhance the productivity of the farmland. In the sub-Saharan region of Africa, the *Brachiaria* grass is an important source of fodder and constitutes a prominent use in pest management strategies. *Brachiaria* genotypes are used as a "pull" component for cereal pests in the climate-adapted push-pull technology (PPT), a habitat management strategy developed to manage the lepidopterous stem borers and spider mites [6]. A reduction in the pest population can be accomplished by recognizing and identifying their feeding preferences—the more pronounced the feeding preferences, the greater the reduction in the population [9].

Consequently, the damage to host plants grown in mixed sowing systems is considerably reduced. Monophagous insects are specific in this regard. The slight alteration of a host plant's canopy renders monophagous insects unable to locate an adequate food supply and establish a suitable breeding base. A significant reduction in the population of oligophagous insects (insects whose host spectrum is in the botanical family) is expected to occur in mixtures of adequately spaced botanical taxa. For example, cereal plants' damage can be reduced by introducing the cereal leaf beetle to cereal–legume mixtures [9]. Ecological aspects of mixed related to the limitation of herbivores favor a green deal in Europe.

A mixture of two or more plants could be introduced to increase domestic protein sources in feed and reduce the protein sources of GM feed. Yellow lupine is an alternative for GM soybean, and its use in forage will depend on the direction of economic activity. Strip intercropping with yellow lupine, a crop of low competitiveness and high sensitivity to other crops' proximity, proved to be the best solution, along with growing triticale as a companion crop with a path separating both species [3]. A proper selection of species for the inter-cropping is a key factor for their optimal development and yield. For example, in the temperate climate of Germany, some species, i.e., alfalfa, sweet yellow clover, and common vetch, proved to be unsuitable for row-intercropping with maize due to difficulties in weed control or allelopathic effects [7]. Components used for interspecies mixtures should have attributes such as uniformity of growth rate and time to mature so that the harvest date may be at the same time.

Moreover, none of the species should be too aggressive, especially when mixed with a more valuable protein plant. In the case of well-established mixtures, e.g., oats–common vetch, where cereal component is more competitive toward the legume one, a proper selection of oat cultivars may also significantly affect the quantity and quality of the mixture's yield, i.e., protein content [1]. The oats–common vetch mixtures develop higher LAI and give a higher seed yield in the conventional farming system; however, the share of vetch seeds in the mixtures is higher in the organic system than in the conventional one [2].

The water content in the soil is a factor that intensifies the inter-species competition, as was shown for barley, undersown with rye-grass, which further weakens the phosphorus uptake by both barley and rye-grass [4]. Water deficit in the soil resulted in barley being a stronger competitor, with rye-grass for phosphorus. The barley competition proved to be a stronger factor hindering phosphorus accumulation in the stems and leaves than water deficit. The strongest competition was noted at the most intense stages of barley development, i.e., during the stem elongation and heading.

The grain quality of spring cereal mixtures was also raised in [8], who found higher protein yields in mixtures of barley (hulled or naked grains) with wheat and the highest yields of net metabolic energy in a mixture of naked barley with wheat. Productive and ecological aspects of mixed cropping systems should be considered when recommending these cultures in climate-smart agriculture [4].

Conflicts of Interest: The authors declare no conflict of interest.

References

1. Pużyńska, K.; Pużyński, S.; Synowiec, A.; Bocianowski, J.; Lepiarczyk, A. Grain Yield and Total Protein Content of Organically Grown Oats–Vetch Mixtures Depending on Soil Type and Oats' Cultivar. *Agriculture* **2021**, *11*, 79. [CrossRef]
2. Pużyńska, K.; Synowiec, A.; Pużyński, S.; Bocianowski, J.; Klima, K.; Lepiarczyk, A. The Performance of Oat-Vetch Mixtures in Organic and Conventional Farming Systems. *Agriculture* **2021**, *11*, 332. [CrossRef]
3. Gałęzewski, L.; Jaskulska, I.; Wilczewski, E.; Wenda-Piesik, A. Response of Yellow Lupine to the Proximity of Other Plants and Unplanted Path in Strip Intercropping. *Agriculture* **2020**, *10*, 285. [CrossRef]
4. Kostrzewska, M.K.; Jastrzębska, M.; Treder, K.; Wanic, M. Phosphorus in Spring Barley and Italian Rye-Grass Biomass as an Effect of Inter-Species Interactions under Water Deficit. *Agriculture* **2020**, *10*, 329. [CrossRef]
5. Klima, K.; Synowiec, A.; Puła, J.; Chowaniak, M.; Pużyńska, K.; Gala-Czekaj, D.; Kliszcz, A.; Galbas, P.; Jop, B.; Dąbkowska, T.; et al. Long-Term Productive, Competitive, and Economic Aspects of Spring Cereal Mixtures in Integrated and Organic Crop Rotations. *Agriculture* **2020**, *10*, 231. [CrossRef]
6. Cheruiyot, D.; Midega, C.A.O.; Pittchar, J.O.; Pickett, J.A.; Khan, Z.R. Farmers' Perception and Evaluation of *Brachiaria* Grass (*Brachiaria* spp.) Genotypes for Smallholder Cereal-Livestock Production in East Africa. *Agriculture* **2020**, *10*, 268. [CrossRef]

7. Schulz, V.S.; Schumann, C.; Weisenburger, S.; Müller-Lindenlauf, M.; Stolzenburg, K.; Möller, K. Row-Intercropping Maize (Zea mays L.) with Biodiversity-Enhancing Flowering-Partners—Effect on Plant Growth, Silage Yield, and Composition of Harvest Material. *Agriculture* **2020**, *10*, 524. [CrossRef]
8. Leszczyńska, D.; Klimek-Kopyra, A.; Patkowski, K. Evaluation of the Productivity of New Spring Cereal Mixture to Optimize Cultivation under Different Soil Conditions. *Agriculture* **2020**, *10*, 344. [CrossRef]
9. Wenda-Piesik, A.; Piesik, D. Diversity of Species and the Occurrence and Development of a Specialized Pest Population—A Review Article. *Agriculture* **2021**, *11*, 16. [CrossRef]

 *agriculture*

Article

Long-Term Productive, Competitive, and Economic Aspects of Spring Cereal Mixtures in Integrated and Organic Crop Rotations

Kazimierz Klima *, Agnieszka Synowiec *, Joanna Puła, Maciej Chowaniak, Katarzyna Pużyńska, Dorota Gala-Czekaj, Angelika Kliszcz, Patryk Galbas, Beata Jop, Teresa Dąbkowska and Andrzej Lepiarczyk *

Department of Agroecology and Crop Production, Faculty of Agriculture and Economics, University of Agriculture in Krakow, Al. Mickiewicza 21, 31-120 Krakow, Poland; joanna.pula@urk.edu.pl (J.P.); maciek.chowaniak@gmail.com (M.C.); k.puzynska@urk.edu.pl (K.P.); d.gala@urk.edu.pl (D.G.-C.); angelika.kliszcz@student.urk.edu.pl (A.K.); patryk.galbas@urk.edu.pl (P.G.); beata.jop@urk.edu.pl (B.J.); t.dabkowska@urk.edu.pl (T.D.)
* Correspondence: rrklima@cyf-kr.edu.pl (K.K.); a.synowiec@urk.edu.pl (A.S.); andrzej.lepiarczyk@urk.edu.pl (A.L.)

Received: 21 May 2020; Accepted: 12 June 2020; Published: 15 June 2020

Abstract: Cultivation of spring cereal mixtures (SCMs) is one of the ways to increase the yield of crops in mountainous areas of Poland. There are only a few current long-term studies on this topic. Our study aimed at analyzing yield and competitiveness as well as the economic indicators of spring cereals in pure or mixed sowings in integrated or organic crop rotations over nine years. A field experiment including pure sowings of oats, spring barley, or spring triticale and their two-component SCMs, each in two systems, organic and integrated crop rotation, was carried out in the Mountainous Experimental Station in Czyrna, Poland, in the years 2011–2019. On average, cereals in the pure sowings and mixtures yielded 18% lower in the organic rotations compared with the integrated ones. However, SCMs yielded higher than the pure sowings, and displayed a higher leaf area index and land equivalent ratio. The average gross margin without subsidies was almost two times higher in the organic crop rotations than in the integrated ones, which was influenced mainly by the cultivation of barley in pure sowing. Summing up, the cultivation of SCMs in the mountainous areas of southern Poland is advised because of both productive and economic factors.

Keywords: barley; oats; triticale; yield; leaf area index; land equivalent ratio; standard gross margin

1. Introduction

The cultivation of spring cereal mixtures (SCMs) is an element of crop rotation typical for Polish agriculture [1]. In 2015, cereals accounted for 73.3% of the total crop area of Poland, including 10.7% of SCMs. In 2017 and 2018, cereals accounted for 70.1% and 72.1%, respectively, and in the same years, the SCMs accounted for 11.6% and 12.7%, respectively [2].

The spring cereal mixtures are applied both in organic and sustainable agricultural systems [3–5], mainly as a source of feed (grains) for livestock [6–8]. This method of crop cultivation involves the simultaneous sowing of usually two different species of spring cereals, in different proportions [9,10]; their grains are mixed before sowing.

Cultivation of SCMs has many advantages and is desirable in sustainable agriculture owing to current ecological trends related to reducing the amount of mineral fertilizers and pesticides, as a part of integrated pest management [11–13]. It results in less environmental pollution and lower outlays on agronomic practices [14]. This is because of the fact that SCMs are less infested by pathogens and insect pests than pure sowings [15,16], which results from a reduced number of plants of a given species susceptible to a particular pest [17]. The components of mixtures are characterized by a different growth pattern [10], and as a result, they better cover soil, protecting it from water loss. This also results in a reduced weed infestation, as species in the mixture compete more effectively with weeds [18] and promote biodiversity in a canopy [19].

Many authors [20–22] confirmed the relationship between grain yield and leaf area index (LAI). LAI is referred to as the ratio of the surface of assimilation organs of a crop, mainly leaves, to the surface of soil. The LAI value depends on genetic characteristics and habitat factors [23]. For various crops, the LAI is several times larger than the surface of the soil, for example, for small-seeded legumes, 4–5 times, and for other crops, 2–4 times [24].

The SCMs are characterized by greater yield stability than pure sowing [25]. This is because of, among others, the fact that, in the adverse weather and soil conditions (e.g., drought) for the first species (component), the second component of the SCM finds more favorable conditions and increases the yield, compensating for the lower yield of the first one [26,27]. This compensation is also associated with the complementary use of soil resources, that is, nutrients and water, which results from the diverse architecture of the root systems of SCMs. Mixtures are an important element in increasing species diversity in crop rotation [28,29], which is supportive to stabilizing the yield of following crops [30].

A common component of SCMs is oats, called a phytosanitary plant [31]. Oats are a cereal species with a very well-developed root system, capable of taking water and nutrients from deeper layers of soil. Moreover, as an allelopathic crop, they influence, by root exudates, the abundance and composition of soil-pests, as well as the composition of soil microorganisms, which together contribute to the soil biological activity [32]. In this way, oats also stimulate the growth and development of the other component of SCM [33].

The available literature lacks current studies on the yield, competitiveness of components, and economic aspects of the cultivation of SCMs in different cropping systems, especially in the long term and in the extensive conditions of mountainous agriculture. Decisions related to the selection of the production structure are made based on both the production/quality characteristics and economic results [34], also including the system of agricultural subsidies [35]. Moreover, subsidies in the mountainous areas of southern Poland are one of the most important and motivating factors for a farmer to produce organically [3]. All of this became a reason for undertaking our research. The profitability of cultivation is determined by the relation between the value of the obtained crop and the incurred production costs, which include all elements throughout the production process. For this reason, our research also included economic analysis of standard gross margin of SCMs' cultivation.

The aims of our study were to (i) analyze the yield, competitiveness, and leaf area index (LAI); and (ii) assess the economic indicators of spring cereals in pure or mixed sowings in integrated or organic crop rotations, over nine years (three rotations of crops) in the mountainous area of southern Poland.

2. Materials and Methods

A field experiment was carried out in the years 2011–2019 in the Mountainous Experimental Station in Czyrna near Krynica Górska, southern Poland (545 m a.s.l.; 49°25′, N 20°58′ E). The soil was acid (pH_{KCl} = 5.1), brown soil—Cambisol [36], formed from weathered flysch material, composed of loam with a medium skeleton content. The chemical composition of soil was as follows: 0.22% $N_{tot.}$; 46.2 mg kg^{-1} soil P; 203.3 mg kg^{-1} soil K; 1.84% C_{org}. The experiment was set up in a two-factorial split-block design, with four replications. The total area of a single plot was of 30.8 m^2, with 22 m^2 of a harvested area. Three full rotations of the crops (nine years) were included in the results.

There were two systems (first factor) of the experiment: (1) integrated, with mineral fertilization and chemical pesticides; and (2) organic, without any synthetic additives. Each of the systems was composed of six three-field crop rotations (second factor): (1) potato fertilized with manure (33 t ha^{-1}); (2) spring cereal pure sowing or a spring cereals mixture—six variants in total (Table 1); and (3) spring vetch. The density of cereals in the mixture was reduced by 50%, in relation to the pure sowings. All of the systems and crops were present each year, which means that each crop was grown nine times throughout the whole study period. Three full rotations of the crops (nine years) were included in the results.

Table 1. Species composition and number of grains (pcs. m^{-2}) for pure and mixed sowings of spring cereals.

Species	Number of Germinating Grains
Pure sowing [1]:	
Oats cv. Borowiak	650
Spring barley cv. Boss	410
Spring triticale cv. Wanad	568
Mixture:	
Oats + spring barley	325 + 205
Oats + spring triticale	325 + 284
Spring barley + spring triticale	205 + 284

[1] Breeders: oats, Małopolska Hodowla Roślin-HBP sp. z o.o. (Krakow, Poland); spring barley, Hodowla Roślin (HR) Smolice Sp. z o. o. Grupa Instytut Hodowli i Aklimatyzacji Roślin (IHAR) (Smolice, Poland); spring triticale, HR Strzelce Sp. z o. o. Grupa IHAR (Strzelce, Poland).

In the integrated crop rotation, a mineral fertilization for cereals in pure sowings and their mixtures was balanced, based on the content of nutrients in soil, quality of the preceding crop, and forecasted yield. In autumn, 34 kg ha^{-1} P and 55.6 kg ha^{-1} K were applied before a deep ploughing in October. For cereals in spring, a total dose of 72.0 kg ha^{-1} N was divided into two equal doses, one applied before sowing and a second in the shoot formation. Grains were coated with karboxine + thiram (60 g + 60 g per 100 kg of grains). Weeds in the pure sowings and mixtures were controlled by tribenuron methyl (12 g ha^{-1}).

In the organic crop rotation, no chemical fertilizers nor pesticides were applied. Weeds in the cereals were mechanically controlled by a Weeder harrow, run two times in by the end of tillering/beginning of shooting (BBCH 29–30 [37]).

In the stage of grains development (BBCH 70–71), samples of cereals were collected from each plot from 1 m^2. An area of leaves was measured from 20 shoots per sample, using an LI-COR 3100 Area Meter (LI-COR Biosciences GmbH, Bad Homburg vor der Höhe, Germany). Next, the average area of leaves per shoot was multiplied by the total number of shoots per 1m^2 [24]. On this basis, a leaf area index (LAI) was calculated:

$$\text{LAI} = \text{Leaf area (m}^2\text{) / Ground cover (m}^2\text{),} \tag{1}$$

At harvest, the grain was collected from each plot (22 m^2). The yield was expressed as per ha at 15% of seed moisture content.

To assess the competition between the components of the mixtures, two competition indices were calculated: land equivalent ratio (LER) (2) [38] and competitive ratio (CR) (3) [39].

$$\text{LER} = \text{LER}i + \text{LER}j, \tag{2}$$

$$\text{LER}i = Yij/Yii, \tag{3}$$

$$\text{LER}j = Yji/Yjj, \tag{4}$$

where Yii—yield of species i in a pure sowing, Yjj—yield of species j in a pure sowing, Yij—yield of species i in a mixed sowing with a species j, and Yji—yield of species j in a mixed sowing with a species i.

If LER value is greater than one (LER > 1), it means that the mixture is more effective than the pure sowing [40,41].

$$CR_i = (LER_i / LER_j)\,(Z_{ji} / Z_{ij}), \tag{5}$$

$$CR_j = (LER_j / LER_i)\,(Z_{ij} / Z_{ji}), \tag{6}$$

where Z_{ij}—proportion of species i in the mixture with species j and Z_{ji}—proportion of species j in the mixture with species i.

If CR = 1, it means that there are equal competitive abilities of species i and j. If CR_i > 1, it means that species i is more competitive than species j. If CR_i < 1, it means that species j is more competitive than species i [40–42].

The economic indicators were calculated. The amount of cash outlay on the means of production was taken as the basis for the agricultural techniques used in the experiment, as well as the consumption of pesticides, fertilizers, and seeds. The values were calculated per area of 1 hectare. The commodity value of the harvested crops and the prices of the means of production were according to data contained in the market analyses developed at the Department of Market Research IERiGZ-PIB, Warsaw, Poland [43]. An additional source of data was the farm production calculations compiled by the Department of Economics and Agricultural Management MODR, Karniowice, Poland [44]. All calculations took into account the prices of the last, that is, the year 2019. The average yield of crops during the 2011–2019 study period was included in the calculations. The amount of human labor expenditure was adopted [45]. The costs of agrotechnical operations were determined using the method [46]. The standard gross margin was calculated from the difference between the value of products obtained and the direct costs incurred. The direct profitability index, which characterizes the relation of production value to direct costs, was determined [47]. The labor consumption in the cultivation of cereals and cereal mixtures was 9.5 working hours per ha^{-1} and was based on the workload involved in the experiment.

The mean value of nine years was used in the statistical analysis. Before the examination, the results of the experiment were tested for normality of distribution as well as homogeneity of variance by Shapiro–Wilk and Brown–Forsythe tests (Statistica PL ver. 13.1, StatSoft, Krakow, Poland). Both tests turned insignificant values (p = 0.1 for both tests), a basis for performing the analysis of variance (ANOVA). The results were subjected to a two-factor ANOVA for a split-block design with four replications, using FR-ANALWAR-4.3 Microsoft Excel-based package (author: Prof. F. Rudnicki; UTP University of Science and Technology, Bydgoszcz, Poland). The significance of differences between means was tested using the Tukey test ($p \leq 0.05$).

Weather Conditions

The weather data were collected from the meteorological station located in the Mountainous Experimental Station in Czyrna near Krynica Górska. The average precipitation in January, April, July, and August for the years 2011–2019 was similar for the multi-year (1962–1990) period. On average, the precipitation in May 2011–2019 was ca. 17 mm more, and in June 2011–2019 was ca. 24 mm less, than for the multi-year period (Table 2).

The average temperature for the vegetative period (April–August) of the 2011–2019 years was 2.4 °C higher as compared with a similar period of a multi-year (Table 3). On average, April and August of the 2011–2019 years were particularly warmer than for a multi-year, by 2.8 and 3.4 °C, respectively. Moreover, the average temperatures in January and May–July of 2011–2019 were ca. 2 °C higher from those of the multi-year period (Table 3).

Table 2. Sum of precipitation (mm) during the course of study.

Year	Month												Ap.–Ag.	J.–Dc.
	J.	Fb.	Mr.	Ap.	M.	Jn.	Jl.	Ag.	Sp.	Oc.	Nv.	Dc.		
2011	36.7	15.1	27.6	106.3	72.1	44.4	278.4	85.6	15.9	34	11.1	15	586.8	742.2
2012	60.9	33.2	20.5	56.6	20.6	167.7	82.2	63.3	45.4	108	25.8	31.6	390.4	715.8
2013	74.1	26.6	38.4	24.7	118	202.4	33.1	32.9	109.6	18	92.9	29.7	411.1	800.4
2014	45.8	21.2	39.2	51.1	137.8	58.3	134.4	113.6	72.7	39.6	45.2	47.5	495.2	806.4
2015	58.8	32.2	45.4	50.5	123.8	43.5	52.1	83.7	82.5	51.3	43.8	52.1	353.6	719.7
2016	21.6	73.8	36.6	62.4	56.2	62	173.1	116.9	55.9	140.5	49.7	53.6	470.6	902.3
2017	162.5	81.2	45.4	121.6	69.1	38.5	100.3	89.7	189.8	59.4	53.6	42.1	419.2	1053.2
2018	67.4	11.8	16.5	25.1	82.4	85.2	118.6	85.4	85.1	53.7	48.5	42.3	396.7	722
2019	51.4	15.8	25.4	85.3	234.2	26.7	60.6	94	89.7	42.5	39.1	51.6	500.8	816.3
2011–2019	**64.4**	**34.5**	**32.8**	**64.8**	**101.6**	**81**	**114.8**	**85**	**83**	**60.8**	**45.5**	**40.6**	**447.2**	**808.7**
1961–1990 [1]	**58.1**	**46.9**	**48**	**62.2**	**84.9**	**105**	**114.9**	**98.3**	**78.7**	**56**	**43.9**	**51.2**	**465.3**	**848.1**

[1] Multi-year period 1961–1990.

Table 3. Mean temperatures (°C) during the course of study.

Year	Month												Ap.–Ag.	J.–Dc.
	J.	Fb.	Mr.	Ap.	M.	Jn.	Jl.	Ag.	Sp.	Oc.	Nv.	Dc.		
2011	−2.5	−4	2.2	8.9	12.3	17.1	16.3	17.9	12.6	7.2	1.6	−0.1	14.5	7.47
2012	−2.6	−7.9	3.1	8.2	13.8	16.2	18.9	17.7	12.7	7.3	3.7	−3.8	14.9	7.27
2013	−3.7	−1.6	−2.1	7.2	13.1	15.5	18.1	17.7	11.4	8.7	3.6	−0.1	14.3	7.3
2014	−3	0.7	5.3	8.5	12.6	14.4	18.7	16.2	13.9	8.8	3.7	−0.9	14	8.32
2015	−0.1	1	3.1	7	11.6	15.7	18.9	20.2	13.2	7.8	1.1	−2.4	14.6	7.92
2016	−3.6	2.5	3.5	7.8	12.8	17.5	18.1	16.6	14.2	6.8	1	−2.5	14.5	7.9
2017	−6.8	−1.1	4.9	6.2	12.3	16.8	17.7	18.4	11.6	7.3	1.4	−2.4	14.2	7.19
2018	−0.3	−4.4	−0.7	12.5	15.8	17.1	18.6	19.3	12.1	7.1	1.5	−2.2	16.6	8.01
2019	−1.6	−3.8	0.8	14.9	16	18.1	19.3	20.2	12.1	8.1	1.2	−1.8	17.7	8.62
2011–2019	**−2.7**	**−2.1**	**2.2**	**9**	**13.4**	**16.5**	**18.3**	**18.2**	**12.6**	**7.7**	**2.1**	**−1.8**	**15**	**7.8**
1961–1990 [1]	**−4.4**	**−3.2**	**1.2**	**6.2**	**11.5**	**14.2**	**16**	**14.8**	**11.2**	**7**	**0.9**	**−2.7**	**12.6**	**6.1**

[1] Multi-year period 1961–1990.

3. Results

3.1. Yield of Cereals in Pure Sowings and Mixtures

The average yield of grain of spring cereals in the organic crop rotation was 18% lower than in the integrated one (Table 4). On average, spring cereal mixtures (SCMs) yielded 8.5% higher than pure sowings. For pure sowings, the lowest was the yield of oats, by 13% less than that of spring barley. Among SCMs, the highest yield was for oats and barley, by 22.6% more than for oats in pure sowing. Moreover, the average yield of a mixture of oats and triticale was 8.4% higher than the yield of oats in pure sowing. The yield of a mixture of triticale and barley was similar to that of a barley in pure sowing, but 11.4% higher than triticale in pure sowing (Table 4).

Table 4. Grain yield (t ha^{-1}) of spring cereals grown in pure sowings or in mixtures in integrated or organic crop rotation, means for the years 2011–2019.

Cereal/Cereal Mixture	Crop Rotation		Mean [1]
	Integrated	Organic	
Oats	4.14	3.21	3.67 A
Spring barley	4.58	3.73	4.15 C
Spring triticale	4.18	3.35	3.76 AB
Oats + spring barley	4.86	4.15	4.50 D
Oats + spring triticale	4.37	3.6	3.98 BC
Spring triticale + spring barley	4.55	3.84	4.19 C
Mean	4.44 B	3.64 A	

[1] Means with various letters are significantly different, according to Tukey test ($p \leq 0.05$).

The grain yield of individual components of SCMs was significantly differentiated (Table 5). On average, grain yields in the integrated crop rotations were 19% higher than in the organic ones. The lowest yield was for the oats in the mixture with triticale in the organic crop rotation. In turn, the barley mixed with oats in the integrated crop rotation yielded the highest (Table 5).

Table 5. Grain yields (t ha^{-1}) of the components of the spring cereal mixtures in the integrated or organic crop rotations, means for the years 2011–2019.

Component of Mixture	Crop Rotation		Mean [1]
	Integrated	Organic	
Oats +	2.24 ef	1.86 b	2.05 A
spring barley	2.62 h	2.29 f	2.45 C
Oats +	2.15 d	1.74 a	1.94 A
spring triticale	2.22 e	1.86 b	2.04 A
Spring barley +	2.35 g	1.97 c	2.16 B
spring triticale	2.2 de	1.87 b	2.03 A
Mean [1]	2.29 B	1.93 A	

[1] Means/values with various letters are significantly different, according to Tukey test ($p \leq 0.05$).

3.2. Leaf Area Index of Cereals in Pure Sowings and Mixtures

The data presented in Table 6 show that the distribution of average values of leaf area index (LAI) for both crop rotations, as well as pure sowings and SCMs, was similar to those for the grain yield (Table 3). The mean value of LAI for the integrated crop rotations was significantly higher than for the organic ones. The values of LAI for SCMs were 9.7% higher than for pure sowings. The highest LAI was recorded for a mixture of oats and barley; it was higher than that of a pure sowing of oats and pure sowing of barley, by 28% and 8%, respectively.

Table 6. Leaf area index ($m^2\ m^{-2}$) of cereals in pure sowings or in mixtures in integrated or organic crop rotations, means for the years 2011–2019.

Cereal/Cereal Mixture	Crop Rotation		Mean [1]
	Integrated	Organic	
Oats	1.93	1.5	1.71 A
Spring barley	2.23	1.82	2.02 BC
Spring triticale	2.07	1.67	1.87 AB
Oats + spring barley	2.36	2.02	2.19 D
Oats + spring triticale	2.12	1.76	1.94 B
Spring triticale + spring barley	2.25	1.9	2.07 CD
Mean [1]	2.16 B	1.77 A	

[1] Means with various letters are significantly different, according to Tukey test ($p \leq 0.05$).

A detailed analysis of LAI for the components of the mixtures revealed that oats in the SCMs had a significantly lower LAI; each time, the value of LAI of oats in the SCMs was below 1. The highest LAI was recorded for barley in the SCMs. What is more, the LAI of barley in the mixture with oats was 30% higher than that of oats (Table 7).

Table 7. Leaf area index ($m^2\ m^{-2}$) of the components of the spring cereal mixtures in integrated or organic crop rotations, means for the years 2011–2019.

Component of Mixture	Crop Rotation		Mean [1]
	Integrated	Organic	
Oats +	1.04	0.86	0.95 B
spring barley	1.32	1.16	1.24 E
Oats +	1.01	0.81	0.91 A
spring triticale	1.11	0.95	1.03 C
Spring barley +	1.15	0.96	1.05 D
spring triticale	1.1	0.94	1.02 C
Mean [1]	1.12 B	0.94 A	

[1] Means with various letters are significantly different, according to Tukey test ($p \leq 0.05$).

3.3. Competition Indices for the Mixtures

Expressing the cereal grain yields as a land equivalent ratio (LER), which shows the productivity of the SCMs, it was found that the LER values for all the SCMs were higher than 1 (Table 8). An SCM of oats with barley, followed by a mixture of oats with triticale and a mixture of barley with triticale, had the highest yielding potential, that is, the highest LER. Moreover, the system of crop rotation, integrated or organic, significantly differentiated the LER value, which was on average 4% higher in the organic system.

Table 8. Values of land equivalent ratio (LER) for the components of the spring cereal mixtures in integrated or organic crop rotations, means for the years 2011–2019.

Component of Mixture	Crop Rotation					
	Integrated		Organic		Mean	
	Component of Mixture	Sum [1]	Component of Mixture	Sum [1]	Component of Mixture	Sum [1]
Oats +	0.54		0.57		0.55	
spring barley	0.57	1.11 B	0.61	1.18 B	0.59	1.14 B
Oats +	0.52		0.54		0.53	
spring triticale	0.53	1.05 A	0.55	1.09 A	0.54	1.07 A
Spring barley +	0.51		0.53		0.52	
spring triticale	0.52	1.03 A	0.55	1.08 A	0.53	1.05 A
Mean [1]		1.06 a		1.11 b		

[1] Means/sums with various letters are significantly different, according to Tukey test ($p \leq 0.05$).

On contrary, the competitive ratio (CR) for the integrated and organic crop rotations was similar (Table 9). For this indicator, both the competitor's species and the component species in the SCM were key. Thus, triticale had the highest CR in the SCMs. Barley was a competitor to oats but underwent competitive pressure when mixed with triticale. Oats undergo competitive pressure in all of the SCMs, with spring barley being the strongest competitor to it.

Table 9. Values of competitive ratio (CR) of the components of the spring cereal mixtures, means for the years 2011–2019.

Component of Mixture	Crop Rotation		Mean [1]
	Integrated	Organic	
Oats +	0.94	0.93	0.93 A
spring barley	1.05	1.07	1.06 C
Oats +	0.98	0.98	0.98 B
spring triticale	1.02	1.02	1.02 BC
Spring barley +	0.98	0.96	0.97 AB
spring triticale	1.02	1.04	1.03 C
Mean [1]	0.99	1	

[1] Means with various letters are significantly different, according to Tukey test ($p \leq 0.05$).

3.4. Economic Indices for the Mixtures

The results presented in Table 10 relate to research carried out in the mountainous area of southern Poland. The basic function of these areas is protection and retention of water resources. Table 10 contains economic measures for the cultivation of spring cereals and spring cereals mixtures. The average direct costs in the integrated system (EUR 481.9 ha^{-1}) were 39.1% higher than in the organic system (EUR 293.5 ha^{-1}). The highest share in the direct costs in the integrated system can be attributed to the application of mineral fertilizers, EUR 138.5 ha^{-1} on average.

Table 10. Economic indicators of pure and mixed sowings of spring cereals in integrated or organic crop rotations (EUR ha^{-1}); prices for the year 2019.

Source of Cost	Pure Sowing						Spring Cereal Mixture						Mean	
	Oats		Barley		Triticale		Oats + Barley		Oats + Triticale		Triticale + Barley		I	O
	I	O	I	O	I	O	I	O	I	O	I	O		
Input costs	544.3	421.9	612.8	499.2	588.9	471.9	570.6	496.8	523.1	431.1	544.7	459.7	564.0	463.3
SGM without subsidies	57.7	123.7	138.6	214.4	103.7	175.1	89.9	204.7	37.2	133.6	64.9	168.3	82.1	169.8
SGM with subsidies	235.6	487.6	316.5	578.3	281.6	539.0	267.8	568.6	215.1	497.5	242.8	532.2	260.0	533.7
Share of subsidies in the SGM (%)	75	74	56	63	63	67	66	64	82	73	73	68	68	68
Direct profitability index														
Without subsidies	1.11	1.41	1.29	1.74	1.21	1.58	1.18	1.69	1.07	1.44	1.13	1.57	1.17	1.57
With subsidies	1.48	2.63	1.66	3.01	1.57	2.81	1.55	2.94	1.44	2.67	1.50	2.82	1.53	2.81

I—integrated; O—organic; SGM—standard gross margin. Conversion rate of 1 EUR = 4.2585 PLN in accordance with the National Polish Bank exchange rate on 31 December 2019.

The average value of direct surplus without subsidies was almost twofold higher in the organic system than in the integrated one. A greater disproportion between the examined systems occurred, when subsidies were added to the direct surplus (Table 10). The sum of subsidies for both cereals in pure sowing and mixtures is the same, and equal to EUR 177.9 ha^{-1} and EUR 363.9 ha^{-1} in the integrated and organic system, respectively.

The highest value of direct surplus without subsidies, among spring cereals, was obtained for spring barley (Table 10). This was because of the fact that pure sowings of spring barley yielded highest. The SCM of oats with barley also yielded high; however, the purchase price of the SCM grains was on average 10% lower than that of spring barley grains.

4. Discussion

Our study presents the results of a nine-year-long field experiment with six different crop rotations, each in two systems—integrated and organic. The results indicated that the average yield of spring cereal grain obtained during the next nine growing seasons in the organic crop rotation was about 18% lower than in the integrated one. This mainly results from the lack of use of easily absorbable fertilizers, as well as pesticides in the organic system. As shown by Kumar et al. [48], at low nutrient availability, especially during early growth of cereals, higher investments in root system development can significantly trade off with aboveground productivity, and strong competition can further strengthen such effects.

It was also found that, in both systems, the yielding of two-species spring cereal mixtures (SCMs) was higher than the cereals in pure sowing, which is consistent with the results of other authors [26,27]. This phenomenon consists of a number of factors, including the complementary use of habitat resources [49] or mutually stimulating allelopathic effect of cereals [50].

Of the three different SCMs, oats with spring barley, oats with spring triticale, and spring triticale with spring barley, higher yields were observed for the mixtures with barley as a component. A mixture of oats and barley was characterized by a particularly high grain yield. Barley was the dominant component in this mixture, which posed a strong competitive effect on oats, as indicated by its high competitiveness ratio (CR = 1.06). Moreover, the barley yield in the mixture with oats was high, and even higher than in pure sowing, taking into account that the plant density reduced by half in mixture compared with pure sowing. Despite the dominance of barley in the mixture with oats, both components of this mixture act complementarily. Spring barley is a low cereal, but with a fast growth rate, high tillering, and a short ripening period [51]. On the contrary, oats are a high cereal, ripening relatively late. As pointed by Shaaf et al. [52], faster initial growth of spring barley favors its stronger tillering. This results in a competitive advantage of spring barley over oats in the early stages of growth [53,54]. Sobkowicz [55] points out that, in the phase of emergence, a competition of root systems for soil resources is more important than those of the aboveground parts for light. This applies especially to spring barley, which, in the early phases of growth, produces a large root system [55]. According to Cousens [56], competition for light begins in the tillering phase. The competitive advantage of oats over spring barley begins in the flowering phase. From this phase, plants of oats are higher than spring barley plants. As pointed by Hecht et al. [57], in later growth phases, when barley density is higher, the stem mass fraction increases, while the root mass fraction decreases. In the later growth period (watery ripe, BBCH 71), higher plants of oats develop greater panicles and grains [53]. So-called height convergence may be observed for the mixtures; specifically, a shortening of the long-culmed cereal species and an increase in the length of the short-culmed species. As a result, the mixtures have a decreased lodging and a higher yield [58]. This phenomenon can also be observed for the mixture of oats and barley [59].

In the present study, a high value of the land equivalent ratio (LER), which is an indicator of crop productivity [60], was also found for the SCMs. This is consistent with the results of other authors [61]. Interestingly, the average sum of LERs for the SCMs in the organic system was equal to 1.11 and was significantly higher than for those SCMs in the integrated system. This may indicate a complementary and more effective use by the components of SCMs of limited habitat resources, especially in the organic system. Rudnicki [62] showed that, as soil conditions deteriorate, the SCMs are more effective compared with pure sowing. However, at better and fertilized crop stands, the yields of mixtures are similar to the yields of pure sowing. Among the examined SCMs, the highest LER value was recorded for a mixture of oats with spring barley (LER = 1.14). This result further confirms the complementarity of the components of this mixture.

In the scientific literature so far, there are no detailed results of studies on the leaf area index (LAI) of SCMs. Available studies on LAI of pure cereal sowing show that there is a directly proportional relationship between this trait and grain yield [63]. Our results may partly explain the tendency to obtain higher yields of SCMs in comparison with pure sowing. The results of this study showed that, in both SCMs containing barley, the LAI value was the highest, and the presence of barley affected this result. This likely resulted from the difference in the height of the components of the mixtures. Barley, as a low cereal, develops leaves in the lower layers of the canopy, and effectively uses space for assimilation of photosynthetically active radiation.

Organic farming is perceived as an agricultural system that balances multiple sustainability goals by promoting global food and ecosystem security. Whether organic agriculture can expand is determined by its economic competitiveness with the other agricultural systems [64,65]. In our study, we found that the average value of standard gross margin without subsidies was almost twofold higher in the organic crop rotation than in the integrated one. This result is perspective for an organic system and is in accordance with research by Crowder and Reganold [64]. They examined the financial performance of organic and conventional agriculture by conducting a meta-analysis of a global dataset spanning 55 crops grown on five continents. They found out that, without organic premiums, benefit/cost ratios and net present values of organic agriculture were significantly lower than those for conventional agriculture. However, when actual premiums were applied, organic agriculture was significantly more profitable (22–35%) and had higher benefit/cost ratios (20–24%) than conventional agriculture. The authors conclude that organic agriculture can continue to expand even if premiums decline [64]. On the contrary, Rosa-Schleich et al. [66] underline that the ecological benefits for the farmer were partly insufficient to outbalance economic costs in the short term, even though these practices have the potential to lead to higher and more stable yields, increase profitability, and reduce risks in the long term. Still, ecological-economic performance of organic practices is highly context-dependent [66]. One of the factors that can increase profits and reduce the risks of the organic cropping system is a proper selection of crops [67,68]. In our research, this condition was met by including pure sowing of spring barley in the crop rotation. As shown by Omokanye et al. [8], the profits from cultivating mixtures are variable, and not always higher than those of selected crops in pure sowings. Moreover, a significant role in our research was played by subsidies, which are of the highest importance in both the organic system as well as the mountainous areas of southern Poland, where the less favored conditions occur [35].

5. Conclusions

The results of our long-term research carried out in the mountainous areas of southern Poland revealed that yield of SCMs and leaf area index in pure and mixed sowing in the organic crop rotation was lower than in the integrated one, by 18% and 16%, respectively. At the same time, the average yield of spring cereal mixtures with barley was higher than that of pure cereal sowing. Under the conditions studied, the highest yield was obtained for a mixture of oats and spring barley, which could be partially explained by the higher leaf area index value for this mixture (LAI = 2.19). The yields of the other mixtures (oats with spring triticale and spring triticale with spring barley) were higher than the yield of the one of the components of the mixture in pure sowing. Moreover, the average sum of LERs for mixtures in the organic system was 1.11 and was significantly higher than for mixtures in the integrated system. In mixtures, barley displayed the highest competitiveness ratio. The analysis of the land equivalent ratio (LER) also showed that, under the examined conditions, spring cereal mixtures are a more effective form of cultivation than pure sowings. Each time, LER values for the mixtures exceeded 1. At the same time, despite lower yields of spring cereals in the organic crop rotation, the average value of standard gross margin without subsidies in the organic crop rotation was almost twice as high as in the integrated one. Among pure and mixed sowings, the highest value of standard gross margin without subsidies was found for spring barley cultivation. Summing up, ecological cultivation of spring cereal mixtures, having many pro-environmental values and showing a standard gross margin, should be recommended especially in the mountainous areas of southern Poland.

Author Contributions: Conceptualization, K.K.; methodology, K.K.; validation, A.L.; formal analysis, K.K. and A.S.; investigation, K.K., M.C., and P.G.; resources, A.L.; data curation, K.K., P.G., and A.K.; writing—original draft preparation, A.S., J.P., K.P., and B.J.; writing—review and editing, T.D., A.K., and D.G.-C.; visualization, K.K.,

A.S., and T.D.; supervision, K.K.; project administration, K.K. and A.L.; funding acquisition, K.K. All authors have read and agreed to the published version of the manuscript.

Funding: This research was funded by the Ministry of Science and Higher Education of Republic of Poland, grant number DS 3124.

Conflicts of Interest: The authors declare no conflict of interest.

References

1. Leszczyńska, D. State and conditions of cultivation of grain crops mixtures in Poland. *J. Res. Appl. Agric. Eng.* **2007**, *52*, 105–108.
2. Statistics Poland. *Concise Statistical Yearbook of Poland*; Zakład Wydawnictw Statystycznych: Warsaw, Poland, 2018. Available online: https://stat.gov.pl/files/gfx/portalinformacyjny/pl/defaultaktualnosci/5515/1/19/1/maly_rocznik_statystyczny_polski_2018.pdf (accessed on 20 May 2020).
3. Klima, K.; Łabza, T. Yielding and economic efficiency of oats crop cultivated using pure and mixed sowing stands in organic and conventional farming systems. *Zywn.-Nauk. Technol. Ja.* **2010**, *17*, 141–147.
4. Kaut, A.H.E.E.; Mason, H.E.; Navabi, A.; O'Donovan, J.T.; Spaner, D. Organic and conventional management of mixtures of wheat and spring cereals. *Agron. Sust. Develop.* **2008**, *28*, 363–371. [CrossRef]
5. Sobkowicz, P.; Tendziagolska, E.; Łagocka, A. Response of oat-triticale mixture to post-emergence weed harrowing. *Acta Agric. Scand. B.* **2020**, *70*, 307–317. [CrossRef]
6. Jedel, P.E.; Salmon, D.F. Forage potential of spring and winter cereal mixtures in a short-season growing area. *Agron. J.* **1995**, *87*, 731–736. [CrossRef]
7. Juskiw, P.E.; Helm, J.H.; Salmon, D.F. Forage yield and quality for monocrops and mixtures of small grain cereals. *Crop. Sci.* **2000**, *40*, 138–147. [CrossRef]
8. Omokanye, A.; Lardner, H.; Lekshmi, S.; Jeffrey, L. Forage production, economic performance indicators and beef cattle nutritional suitability of multispecies annual crop mixtures in northwestern Alberta, Canada. *J. Appl. Anim. Res.* **2019**, *47*, 303–313. [CrossRef]
9. Juskiw, P.E.; Helm, J.H.; Salmon, D.F. Competitive ability in mixtures of small grain cereals. *Crop. Sci.* **2000**, *40*, 159–164. [CrossRef]
10. Juskiw, P.E.; Helm, J.H.; Salmon, D.F. Postheading biomass distribution for monocrops and mixtures of small grain cereals. *Crop Sci.* **2000**, *40*, 148–158. [CrossRef]
11. Smithson, J.B.; Lenne´, J.M. Varietal mixtures: A viable strategy for sustainable productivity in subsistence agriculture. *Ann. Appl. Biol.* **1996**, *128*, 127–158. [CrossRef]
12. Wolfe, M.S.; Baresel, J.P.; Desclaux, D.; Goldringer, I.; Hoad, S.; Kovacs, G.; Loeschenberger, F.; Miedaner, T.; Østergard, H.; Lammerts van Bueren, E.T. Developments in breeding cereals for organic agriculture. *Euphytica* **2008**, *163*, 323–346. [CrossRef]
13. Pappa, V.A.; Rees, R.M.; Walker, R.L.; Baddeley, J.A.; Watson, C.A. Legumes intercropped with spring barley contribute to increased biomass production and carry-over effects. *J. Agric. Sci.* **2011**, *150*, 584–594. [CrossRef]
14. Klima, K.; Stokłosa, A.; Pużyńska, K. Agricultural and economic circumstances of cereal cultivation under differentiated soil and climate conditions. *Zesz. Probl. Postęp. Nauk Rol.* **2011**, *559*, 115–121.
15. Finckh, M.R.; Gacek, E.S.; Goyeau, H.; Lannou, C.; Ueli, M.; Mundt, C.C.; Munk, L.; Nadziak, J.; Newton, A.C.; de Vallavieille-Pope, C.; et al. Cereal variety and species mixtures in practice, with emphasis on disease resistance. *Agronomie* **2000**, *20*, 813–837. [CrossRef]
16. Dambolena, J.S.; Lopez, A.G.; Meriles, J.M.; Rubinstein, H.R.; Zygadlo, J.A. Inhibitory effects of 10 natural phenolic compounds on *Fusarium verticillioides*. A structure-property activity relationship study. *Food Cont.* **2012**, *28*, 163–170. [CrossRef]
17. Finckh, M.R.; Wolfe, M.S. The use of biodiversity to restrict plant diseases and some consequences for farmers and society. In *Ecology in Agriculture*, 1st ed.; Jackson, L.E., Ed.; Academic Press: San Diego, CA, USA, 1997; pp. 199–233.
18. Hiltbrunner, J.; Jeanneret, P.; Liedgens, M.; Stamp, P.; Streit, B. Response of weed communities to legume living mulches in winter wheat. *J. Agron. Crop. Sci.* **2007**, *193*, 93–102. [CrossRef]
19. Rajaniemi, T.K.; Allison, V.J.; Goldberg, D.E. Root competition can cause a decline in diversity with increased productivity. *J. Ecol.* **2003**, *91*, 407–416. [CrossRef]

20. Woźniak, A.; Gontarz, D.; Staniszewski, M. Effect of crop rotation on yielding and leaf area index (LAI) of hard wheat (*Triticum durum* Desf.). *Biuletyn IHAR* **2005**, *237–238*, 13–21.

21. Srinivasan, V.; Kumar, P.; Long, S.P. Decreasing, not increasing, leaf area will raise crop yields under global atmospheric change. *Glob. Chang. Biol.* **2016**, *23*, 1626–1635. [CrossRef]

22. Hirooka, Y.; Homma, K.; Maki, M.; Sekiguchi, K.; Shiraiwa, T.; Yoshida, K. Evaluation of the dynamics of the leaf area index (LAI) of rice in farmer's fields in Vientiane Province, Lao PDR. *J. Agric. Meteorol.* **2017**, *73*, 16–21. [CrossRef]

23. Fang, H.; Baret, F.; Plummer, S.; Schaepman-Strub, G. An overview of global leaf area index (LAI): Methods, products, validation, and applications. *Rev. Geophys.* **2019**, *57*, 739–799. [CrossRef]

24. Oleksy, A.; Szmigiel, A.; Kołodziejczyk, M. Yielding and leaf area development of selected winter wheat cultivars depending on technology level. *Fragm. Agron.* **2009**, *26*, 120–131.

25. Sobkowicz, P.; Tendziagolska, E.; Lejman, A. Performance of multi-component mixtures of spring cereals. Part 1. Yields and yield components. *Acta Sci. Pol. Agric.* **2016**, *15*, 25–35.

26. Klima, K.; Smaczny, M. Yielding and competitiveness of oats and spring vetch depending of cultivations systems and sowing method. *J. Agric. Eng. Res.* **2015**, *60*, 146–149.

27. Klima, K.; Łabza, T.; Lepiarczyk, A. Yielding of spring triticale grown under organic and integrated systems of farming and economic indicators of its production. *J. Agric. Eng. Res.* **2015**, *60*, 142–145.

28. Wanic, M.; Nowicki, J.; Kurowski, T.P. Regeneracja stanowisk w płodozmianach zbożowych poprzez stosowanie siewów mieszanych. *Zesz. Probl. Postęp. Nauk Rol.* **2000**, *470*, 137–143.

29. Klimek-Kopyra, A.; Bacior, M.; Zając, T. Biodiversity as a creator of productivity and interspecific competitiveness of winter cereal species in mixed cropping. *Ecol. Model.* **2017**, *343*, 123–130. [CrossRef]

30. Vilich, V. Crop rotation with pure stands and mixtures of barley and wheat to control stem and root rot diseases. *Crop. Prot.* **1993**, *12*, 373–379. [CrossRef]

31. Atanasova, D.; Maneva, V.; Nedelcheva, T. Effect of predecessors on the productivity and phytosanitary condition of hull-less oats in organic farming. *Agric. Sci. Technol.* **2016**, *8*, 4–7. [CrossRef]

32. Bednarek, W.; Tkaczyk, P.; Dresler, S.; Jawor, E. Relationship between oat yield and some soil properties and nitrogen fertilization. *Acta Agroph.* **2013**, *20*, 29–38.

33. Buczek, J.; Tobiasz-Salach, R.; Bobrecka-Jamro, D. Assessment of yielding and weeding effects of mixed spring cereals. *Zesz. Probl. Postęp. Nauk Rol.* **2007**, *516*, 11–18.

34. Gawęda, D.; Nowak, A.; Haliniarz, M.; Woźniak, A. Yield and economic effectiveness of soybean grown under different cropping systems. *Int. J. Plant Prod* **2020**, 1–11. [CrossRef]

35. Klima, K.; Kliszcz, A.; Puła, J.; Lepiarczyk, A. Yield and profitability of crop production in mountain less favoured areas. *Agronomy* **2020**, *10*, 700. [CrossRef]

36. IUSS Working Group WRB. *World Reference Base for Soil Resources 2014, Update 2015, International Soil Classification System for Naming Soils and Creating Legends for Soil Maps*; FAO: Rome, Italy, 2014; World Soil Resources Reports No. 106; Available online: http://www.fao.org/publications/card/en/c/942e424c-85a9-411d-a739-22d5f8b6cc41/ (accessed on 20 May 2020).

37. Lancashire, P.D.; Bleiholder, H.; Boom, T.V.D.; Langelüddeke, P.; Stauss, R.; Weber, E.; Witzenberger, A. A uniform decimal code for growth stages of crops and weeds. *Ann. Appl. Biol.* **1991**, *119*, 561–601. [CrossRef]

38. Mead, R.; Willey, R.W. The concept of a land equivalent ratio and advantages in yields for intercropping. *Exp. Agric.* **1980**, *16*, 217–228. [CrossRef]

39. Willey, R.W.; Rao, M.R. A competitive ratio for quantifying competition between intercrops. *Exp. Agric.* **1980**, *16*, 117–125. [CrossRef]

40. Tobiasz-Salach, R.; Bobrecka-Jamro, D.; Szpunar-Krok, E. Assessment of the productivity and mutual interactions between spring cereals in mixtures. *Fragm. Agron.* **2011**, *28*, 116–122.

41. Yu, Y.; Stomph, T.J.; Makowski, D.; van der Werf, W. Temporal niche differentiation increases the land equivalent ratio of annual intercrops: A meta-analysis. *Field Crop. Res.* **2015**, *184*, 133–144. [CrossRef]

42. Dhima, K.V.; Lithourgidis, A.S.; Vasilakoglou, I.B.; Dordas, C.A. Competition indices of common vetch and cereal intercrops in two seeding ratio. *Field Crop. Res.* **2007**, *100*, 249–256. [CrossRef]

43. Zalewski, A.; Chrościcki, J.; Oleksiak, T.; Trajner, M. Rynek środków produkcji i usług dla rolnictwa. In *Analizy Rynkowe*, 1st ed.; IERiGŻ: Warszawa, Poland, 2018; Volume 45, pp. 1–45.

44. Pobereźnik, B. *Kalkulacje Produkcji Rolniczej*, 1st ed.; Wydawnictwo MODR: Karniowice, Poland, 2018; p. 29.

45. Klikocka, H.; Głowacka, A.; Juszczak, D. The influence of different soil tillage methods and mineral fertilization on the economic parameter of spring barley. *Fragm. Agron.* **2011**, *28*, 44–54.

46. Muzalewski, A. *Koszty Eksploatacji Maszyn*, 1st ed.; Wydawnictwo IBMER: Warszawa, Poland, 2009; p. 46.

47. Klepacki, B.; Gołębiewska, B. Opłacalność produkcji ziemniaków jadalnych. In *Produkcja i Rynek Ziemniaków Jadalnych*, 1st ed.; Chotkowski, J., Ed.; Wyd. Wieś Jutra: Warszawa, Poland, 2002; pp. 40–48.

48. Kumar, A.; van Duijnen, R.; Delory, B.M.; Reichel, R.; Brüggemann, N.; Temperton, V.M. Barley shoot biomass responds strongly to N:P stoichiometry and intraspecific competition, whereas roots only alter their foraging. *bioRxiv* **2020**, 912352. [CrossRef]

49. Sobkowicz, P.; Podgórska-Lesiak, M. Experiments with crop mixtures: Interactions, designs and interpretation. *Electron. J. Pol. Agric. Univ.* **2007**, *10*, 22.

50. Narwal, S.S. Allelopathic interactions in multiple cropping systems. In *Allelopathy in Ecological Agriculture and Forestry. Proceedings of the III International Congress on Allelopathy in Ecological Agriculture and Forestry, Dharwad, India, 18–21 August 1998*; Narwal, S.S., Hoagland, R.E., Dilday, R.H., Reigosa, M.J., Eds.; Springer: Berlin, Germany, 2000; pp. 141–157.

51. Sobkowicz, P. Interspecies competition in spring cereal mixtures. *Zesz. Nauk. AR Wrocław-Rozpr.* **2003**, *458*, 1–105.

52. Shaaf, S.; Bretani, G.; Biswas, A.; Fontana, I.M.; Rossini, L. Genetics of barley tiller and leaf development. *J. Integr. Plant. Biol.* **2019**, *61*, 226–256. [CrossRef]

53. Szarek, K. The competitiveness of spring cereals, cultivated as mixtures and pure sowings, depending on the sowing density. *Acta Agrar. Silvest. Agrar.* **2008**, *51*, 3–9.

54. Kaczmarek, S.; Matysiak, K.; Krawczyk, R. The effect of wheat, barley, and oat root system interactions, in two-species mixtures. *J. Plant. Prot. Res.* **2013**, *53*, 65–70. [CrossRef]

55. Sobkowicz, P. Above-ground and underground competition between barley and oat in the mixture in the initial period of growth. *Fragm. Agron.* **2001**, *18*, 103–119.

56. Cousens, R.D. Comparative growth of wheat, barley, and annual ryegrass (*Lolium rigidum*) in monoculture and mixture. *Austral. J. Agric. Res.* **1996**, *47*, 449–464. [CrossRef]

57. Hecht, V.L.; Temperton, V.M.; Nagel, K.A.; Raschel, U.; Postma, J.A. Sowing density: A neglected factor fundamentally affecting root distribution and biomass allocation of field grown spring barley (*Hordeum vulgare* L.). *Front. Plant. Sci.* **2016**, *7*, 1–14. [CrossRef]

58. Zając, T.; Oleksy, A.; Stokłosa, A.; Klimek-Kopyra, A.; Styrc, N.; Mazurek, R.; Budzyński, W. Pure sowings versus mixtures of winter cereal species as an effective option for fodder-grain production in temperate zone. *Field Crop. Res.* **2014**, *166*, 152–161. [CrossRef]

59. Stokłosa, A.; Stępnik, K. Development of differentiated in maturity period oats in mixture with spring barley. *Fragm. Agron.* **2009**, *26*, 116–125.

60. Liu, X.; Rahman, T.; Song, C.; Yang, F.; Su, B.; Cui, L.; Bu, W.; Yang, W. Relationships among light distribution, radiation use efficiency and land equivalent ratio in maize-soybean strip intercropping. *Field Crop. Res.* **2018**, *224*, 91–101. [CrossRef]

61. Sobkowicz, P.; Tendziagolska, E.; Lejman, A. Performance of multi-component mixtures of spring cereals. part 2. Competitive hierarchy and yield advantage of mixtures. *Acta Sci. Pol. Agric.* **2017**, *15*, 37–48.

62. Rudnicki, F. Impact of forecrop on yielding of various cereals in farm conditions. *Fragm. Agron.* **2005**, *22*, 172–182.

63. Rachoń, L.; Szumiło, G.; Michałek, W.; Bobryk-Mamczarz, A. Variability of leaf area index (LAI) and photosynthetic active radiation (PAR) depending on the wheat genotype and the intensification of cultivation technology. *Agron. Sci.* **2018**, *73*, 63–71. [CrossRef]

64. Crowder, D.W.; Reganold, J.P. Financial competitiveness of organic agriculture on a global scale. *Proc. Nat. Acad. Sci. USA* **2015**, *112*, 7611–7616. [CrossRef]

65. Seufert, V.; Ramankutty, N. Many shades of gray—The context-dependent performance of organic agriculture. *Sci. Adv.* **2017**, *3*, 1602638. [CrossRef]

66. Rosa-Schleich, J.; Loos, J.; Musshoff, O.; Tscharntke, T. Ecological-economic trade-offs of Diversified Farming Systems—A review. *Ecol. Econ.* **2019**, *160*, 251–263. [CrossRef]

67. White, K.E.; Cavigelli, M.A.; Conklin, A.E.; Rasmann, C. Economic performance of long-term organic and conventional crop rotations in the Mid-Atlantic. *Agron. J.* **2019**, *111*, 1358–1370. [CrossRef]

68. Wieme, R.A.; Carpenter-Boggs, L.A.; Crowder, D.W.; Murphy, K.M.; Reganold, J.P. Agronomic and economic performance of organic forage, quinoa, and grain crop rotations in the Palouse region of the Pacific Northwest, USA. *Agric. Syst.* **2020**, *177*, 102709. [CrossRef]

Article

Farmers' Perception and Evaluation of Brachiaria Grass (*Brachiaria* spp.) Genotypes for Smallholder Cereal-Livestock Production in East Africa

Duncan Cheruiyot [1,2,*], Charles A.O. Midega [1], Jimmy O. Pittchar [1], John A. Pickett [3] and Zeyaur R. Khan [1]

[1]   International Centre of Insect Physiology and Ecology (*icipe*), P.O. Box 30772, Nairobi 00100, Kenya; cmidega@icipe.org (C.A.M.); jpittchar@icipe.org (J.O.P.); zkhan@icipe.org (Z.R.K.)
[2]   Department of Biological Sciences, Faculty of Science and Technology, University of Kabianga, P.O Box 2030, Kericho 20200, Kenya
[3]   School of Chemistry, Cardiff University, Cardiff CF10 3AT, UK; PickettJ4@cardiff.ac.uk
*   Correspondence: duncan.cheruiyot@yahoo.com

Received: 27 May 2020; Accepted: 24 June 2020; Published: 4 July 2020

Abstract: *Brachiaria* (*Urochloa*) is a genus, common name brachiaria, of forage grasses that is increasingly transforming integrated crop-livestock production systems in East Africa. A study was undertaken to (i) assess smallholder farmers' perception on benefits of brachiaria in cereal-livestock production, (ii) identify brachiaria production constraints, and (iii) identify farmer preferred brachiaria genotypes. A multi-stage sampling technique was adopted for sample selection. Data were collected through semi-structured individual questionnaire and focus group discussions (FGDs). The study areas included Bondo, Siaya, Homabay and Mbita sub-counties in Western Kenya and the Lake zone of Tanzania. A total of 223 farmers participated in individual response questionnaires while 80 farmers participated in the FGDs. The respondents considered brachiaria mainly important in management of cereal pests (70.4% of respondents) and as an important fodder (60.8%). The major production constraint perceived by both male and female respondents is attacks by arthropods pests (49.2% and 63%, respectively). Spider smites had been observed on own farms by 50.8% of men and 63.1% of women, while sorghum shoot flies had been observed by 58.1% of men and 67.9% of women. These pests were rated as a moderate to severe problem. Xaraes was the most preferred genotype, followed by Mulato II and Piata. These genotypes are important in developing new crop pest management strategies, such as push-pull, and for relatively rapid improvements in crop management and yield increases, particularly in developing countries.

Keywords: brachiaria; cereal-livestock production; perception; push-pull technology; smallholder farmers

---

1. Introduction

Brachiaria (*Urochloa*) is a genus in Poaceae family commonly called brachiaria and grown for forage in Latin America, Asia, South Pacific, and Australia [1]. A widely grown species *Brachiaria brizantha* represents 85% of cultivated pastures in brazil alone [2]. In Africa, where they originate from, they are natural constituents of grasslands in eastern, central, and southern regions [3]. Brachiaria is recently identified as an ideal fodder that can improve livestock production in eastern Africa. This is due to its adaptability to low fertility areas, arid, and semiarid zones of sub-Saharan Africa [3]. There are several initiatives in the region aimed at promoting cultivation of brachiaria to support the emerging livestock industry [4].

A reduction of fall armyworm, (*Spodoptera frugiperda*) damage was recently observed in push-pull technology (PPT) plots as compared to farmers' practice [5]. Furthermore, desmodium enhances soil integrity. Further, *Brachiaria* cv. Mulato II is an important companion crop used as a trap plant in a push-pull technology (PPT, www.push-pull.net). Developed by International Centre of Insect Physiology and Ecology (*icipe*, Nairobi, Kenya), Rothamsted Research (Harpenden, UK), and national partners, PPT is a conservation agriculture system for integrated pest, weed, and soil fertility management in crop–livestock farming systems [6,7]. The system involves intercropping the main crop, either maize *Zea mays L.* or sorghum *Sorghum bicolor* (L.) Moench with a fodder legume, silverleaf desmodium *Desmodium uncinatum* (Jacq.) DC., and surrounded with Napier grass, *Pennisetum purpureum* [8]. The climate smart variant uses drought tolerant green leaf desmodium, *Desmodium intortum* (Mill.) Urb., and brachiaria *B. brizantha* cv Mulato II as the border crop [9,10]. Desmodium releases chemicals that repel stemborers, while volatile chemicals from Napier grass or brachiaria attract the insects and their natural enemies. Through this chemistry, PPT significantly reduces the infestation of cereal stemborers *Busseola fusca* (Fuller) (Noctuidae) and *Chilo partellus* Swinhoe (Crambidae) [10–13]. Significant through nitrogen fixation improves soil organic content and conserves the soil moisture [14,15]. Root exudates of desmodium cause the abortive germination of a noxious weed striga *Striga haemonthica*, therefore providing additional benefits in weed management [11]. On the other hand, brachiaria is a high-value forage crop that facilitates milk production and diversifies farmers' sources of income. A recent study shows that *B. brizantha* cv Piata has higher content of dry matter, crude protein, and organic matter than Napier grass [16]. Some species of brachiaria reduce emission of nitrous oxide from the soil through biological nitrification inhibition [17,18].

Agriculture is an economic mainstay in most developing countries of the tropics. It is characterized by smallholder mixed crop-livestock systems where rain-fed crops and livestock are raised on the same farm [19]. In sub-Saharan Africa (SSA), smallholder farming is a major source of food production and income, contributing up to 80 percent of food consumed [20]. Crops mostly cultivated in the region are, in order, maize, cassava, rice, sorghum, wheat, and millet, while livestock species include cattle, goats, and sheep [19]. However, production in the region is constrained by climate change related biotic and abiotic constraints, including pests and disease outbreaks, extreme weather conditions, among others. This poses a threat to food security and livelihood in communities dependent on agriculture [21,22]. More than 250,000 smallholder farmers in sub-Saharan Africa have used push-pull technology to manage stemborers, fall armyworm, noxious Striga weeds, and soil fertility, and to generate livestock fodder [23].

The value of brachiaria to African agriculture observed thus far can be optimized by addressing the current and foreseeable production constraints. Yet, the few genotypes that are commercialized in Africa were developed for the Americas and Australia thus a higher risk of pest and disease attacks, coupled with poor adaptability to local environments. There is, remarkably, a wide genetic variation in the genus *Brachiaria* [24] that can be exploited in breeding programs for locally adapted genotypes. Recent studies identified brachiaria genotypes that combine drought tolerance and moderate resistance to spider mites [25,26]. Among these genotypes, some are attractive to oviposition by stemborer moths while being detrimental to the larvae of the pest, thus valuable in push-pull technology [27]. These genotypes could be of value in the improvement of cereal livestock-based livestock productivity in sub-Saharan Africa in the current scenarios of increasing aridification and attacks by invasive pests, such as spider mite (*Oligonychus trichardti*). However, farmers' skills and knowledge can complement scientific research and their contribution through participatory approach is key in validating the potential of such genetic materials. Therefore, this study aimed at (i) assessing smallholder farmers' perception on benefits of brachiaria in cereal-livestock production, (ii) assessing brachiaria production constrains, and (iii) identifying farmer preferred brachiaria genotypes.

2. Materials and Methods

2.1. Study Area

Arid and semi-arid areas of western Kenya and the lake zone in Tanzania were selected because of their importance in cereal-livestock based farming systems. The study areas in Kenya included Homabay, Mbita, Bondo, and Siaya. The lake zone in Tanzania (hereafter referred to Tanzania LZ) included Tarime and Mwanza districts. Rainfall pattern is bi-modal, main season runs from March to August and the short season is from October to January. Farming systems in the regions are predominantly cereal/edible legume integrated with livestock [6]. These areas are historically hot spots for cereal stemborer, and most farmers have widely adopted push-pull technology (PPT) as a management tool for the pest [5]. Further, the areas are characterized by extended periods of drought, which makes them conducive for the invasive spider mite, *O. trichardti*. Spider mite is the most important pest of brachiaria, a companion crop in PPT, especially during drier and hotter regimes [26]. These study areas are therefore ideal for assessment of farmers' experience and preference of brachiaria genotypes for use in cereal-livestock production.

2.2. Demonstration Plots

One site per study area was selected for the establishment of the demonstration plots for six brachiaria genotypes. This was purposely done by ensuring that they occur in different agro-ecologies, as follows; Homabay (Lower Midland 3), Mbita (Lower midland 5), Bondo (Lower midland 4), and Siaya (Lower midland 2) [28]. The lake zone sites in Tanzania included Tarime (high altitude plateau) and Mwanza (medium altitude plains) [29]. Brachiaria genotypes that were planted for evaluation were Piata, Xaraes, Marandu, ILRI 12991, ILRI 14807, and Mulato II (check). The candidate genotypes were selected from previous studies that tested drought tolerance, adaptability to a range of environments, and resistance to spider mite in brachiaria [25,26]. Furthermore, Xaraes, Piata, and Marandu are suitable for egg laying by the lepidopterous stemborer *Chilo partellus* and are, therefore, suitable companion plants in PPT [27]. Each plot measured 5 × 5 m with plant to plant and row to row spacing of 0.5 m. Diammonium phosphate (DAP) was applied as basal fertilizer at a rate of 60 kg/ha, and nitrogen in the form of calcium ammonium nitrate (CAN), at a rate of 60 kg/ha as at top dresser four weeks after planting. The plots were kept weed free by hoe and hand weeding and pesticides were not applied to allow natural infestation and the development of spider mites.

2.3. Sampling Procedures

The selection of respondents to participate in study at specific trial sites followed a multistage sampling procedure. Firstly, farmers who practiced climate smart PPT and were within the study areas were selected for the study. This was done by generating a checklist of all farmers who practice climate smart PPT with help of village elders and frontline extension staff. Thereafter, a semi-structured questionnaire was used to identify willing respondents for participatory evaluation of different brachiaria genotypes grown in the demonstration plots.

2.4. Data Collection

The study used semi-structured questionnaires that were administered through individual interviews and focus group discussions (FGD). The questionnaires were pre-tested before implementation. Individual response questionnaire assessed farmers' socio-economic characteristics (e.g., age, gender, and education), farm characteristics (farm size, tenure system, size of land under brachiaria, and uses of brachiaria). Farmers' perceptions on whether brachiaria was beneficially in controlling cereal pests, for sale, as livestock feed, for soil conservation, etc., was sought. Production challenges, including access to planting materials, planting, crop management, harvesting, and hay making was recorded. The questionnaire also assessed farmer experience with pests and diseases of brachiaria. This was captured by asking the respondents whether they had noticed the infestation of the red spider mites and sorghum shoot flies; severity of infestation; and, what they did to cope with the pests. Rating of the seriousness of the pest was based on a four-point Likert scale, where 0 = no problem, 1 = moderate problem, 2 = severe problem, and 3 = very severe problem. Farmers were asked whether they are aware of any other brachiaria genotypes and whether they had planted them on their farms. Thereafter, farmers evaluated the different brachiaria genotypes in demonstration based on hairlines, leaf size, leaf softness, number of shoot tillers, plant spread, plant height, seed setting, resistance to spider mites, and biomass yield where their responses were based on a scale of 1 (poor), 2 (fair), 3 (good), and 4 (excellent). However, for each trait, the number of the highest score i.e., 4 (excellent) was used to compare the genotypes. Evaluation based on sorghum shoot fly damage was not done, since there was no infestation; this is because the genotypes were raised from root splits. Sorghum shoot flies are known to attack young seedlings, especially when grown from the seed. During the assessment, the genotypes were given numbers instead of their actual names to reduce bias in ranking of popular genotypes. Farmers were finally asked to select the best brachiaria genotype. To back up individual interviews, focus group discussions were conducted. Farmers were encouraged to use a language that they were most familiar with and discussions were led by a member of the research group who spoke their language. Similarly, the discussions covered the benefits of brachiaria in climate smart PPT, production constraints, including important pests (spider mites and sorghum shoot flies), and farmer preference of different brachiaria genotypes. Other aspects that were covered in the FGD included willingness of the farmers to try other brachiaria genotypes on their farms and the criteria used for selecting a candidate genotype.

2.5. Data Analysis

Descriptive and comparative statistics (means, percentages, and cross tabulations) were used in data analysis. Analysis of variance, F-test, and Pearson's product moment correlation coefficient (chi-square test) were used to test for significance of differences in various responses and study areas. Computation was done using statistical package for SPSS version 17 software (IBM Corporation, Armonk, NY, USA).

3. Results

3.1. Farmer Socio-demographics and Farm Characteristics

A total of 223 respondents participated in individual interviews. In general, 49% of the respondents are male. There was no significance ($p = 0.05$) in variation between districts (Table 1). Their age categories varied significantly ($p = 0.01$) between the study areas. The majority were between 41–50 years (31%), followed by 51–60 years (22%), the elderly >61 (19%), 31–40 (15%), and the least being the youth between 20–30 years (8%). Age category 41–50 years formed the highest percentage in all study areas, except in Siaya, where the majority were between 51–60. The lowest in population in all areas were the youth (20–30 years), implying that farming in the region is predominantly practiced by the older and the elderly farmers. Education levels of the respondents varied significantly across the study areas. On average, 50% of the farmers had attained primary education, 31% had secondary education, a few (9%) had post-secondary education, while those with none-formal and no education at all were the least, each comprising of 4% of the respondents. Literacy at post-secondary level was the highest in Bondo (15%), while illiteracy (no education) was highest in Siaya (13%) (Table 1).

Farmers rented an average of 1.5 acres land for farming. This varied significantly across the study areas ranging from one acre (Homabay) to two acres (Mbita). The average farm size owned was 3.5 acres, and it varied significantly ($p = 0.01$) from two acres (Siaya) to 5.2 (Homabay). The average land size under brachiaria as components of push-pull was 0.17 acres, this however did not vary significantly across the study areas (Table 1). Besides the brachiaria that forms a component of push-pull, 22% of farmers planted the grass as pure stands; 32% in Homabay, 3% in Mbita, 19% in Bondo, 39% in Siaya, and 11% in Tanzania. The majority of the farmers (93%) kept livestock that included cattle (improved and local), goats (improved and local), and sheep. The type of livestock mostly kept was local cattle with a mean of 2.9, followed by local goats (1.94), sheep (1.65), improved dairy cattle (0.72), and improved dairy goats (0.56). Variation across study areas was not significant for all animals, except sheep ($p = 0.01$). The highest number of sheep was recorded in Homabay (2.92), while the lowest was in Siaya (0.64).

Table 1. Socio-economic characteristics of the respondents.

Variable	Homabay ($n = 38$)	Mbita ($n = 38$)	Bondo ($n = 67$)	Siaya ($n = 54$)	Tanzania ($n = 26$)	Mean ($n = 45$)	F Value	χ^2 value
Gender (male) (%)	53	45	46	37	62	48.6		
Age category (%)								32.842 **
20–30	18	8	3	6	4	7.8		
31–40	13	24	16	15	8	15.2		
41–50	24	37	30	24	42	31.4		
51–60	18	13	21	31	27	22		
>61	13	16	28	22	16	19		
	86	98	98	98	97	95.4		
Education level (%)								26.515 *
None	3	0	5	13	0	4.2		
None-formal	3	8	4	2	4	4.2		
Primary	50	42	43	50	69	50.8		
Secondary	34	42	34	20	27	31.4		
Post-Secondary	8	8	15	13	0	8.8		
Average land size rented (acres)	1.08	2.05	1.63	0.92	2		2.856 *	
Land size owned (acres)	5.29	2.8	2.67	2.05	4.74		4.59 **	
Size of brachiaria plots in push pull (m^2)	793	634	655	603	734	670	0.329ns	
Brachiaria grown as pure stands	32	3	19	39	11	22.5		21.159 ***
Keeping of livestock on farm	95	100	84	94	100	93		13.963 **
Improved dairy cattle	0.42	0.97	0.49	0.79	1.23	0.72	1.738ns	
Local cattle	3.42	3.37	2.46	2.51	3.65	2.9	1.727ns	
Improved dairy goats	0.86	0.53	0.63	0.57	0	0.56	1.841ns	
Local goats	1.92	2.02	2.08	1.05	3.3	1.94	2.15ns	
Sheep	2.92	2.68	1.4	0.64	1.65	1.74	4.395 **	

* = significant at 0.05, ** = significant at 0.01, *** = significant at 0.001, ns = not significant.

3.2. Benefits of Brachiaria

The uses of brachiaria varied significantly across the study areas, except for those who exchanged the grass for milk (mean = 5%). Being a component of a pest management strategy, approximately 38% (mean) of farmers agreed that brachiaria reduced damage caused by cereal pests (Figure 1). Approximately 36% of the farmers considered the grass as a valuable fodder for livestock. About 29% believed that it controls soil erosion, while a few (17%) also sold the grass, others exchanged it for milk (Figure 1).

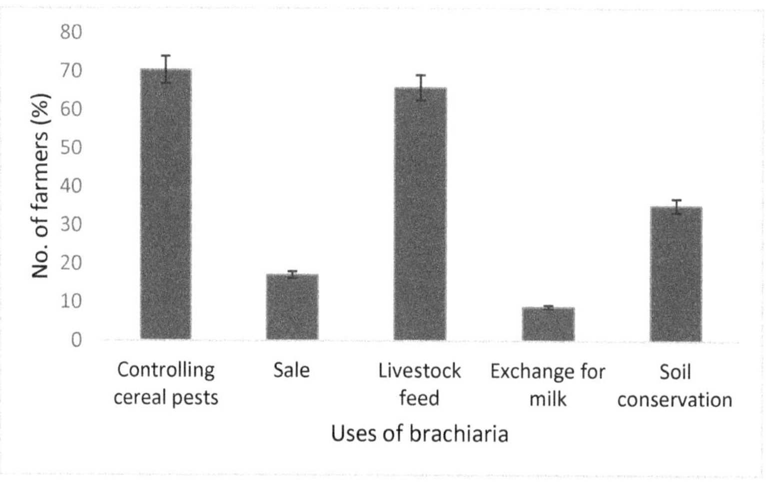

Figure 1. Main uses of brachiaria by the small-holder push pull farmers.

3.3. Constraints to Production of Brachiaria

Farmers gave their opinions on challenges faced in accessing planting materials, during planting and management (Table 2). The opinions varied across the study sites. The major production constraint perceived by both men and women was attacks by arthropod pests (49.2% and 63.1, respectively). This was followed by disease infestation (44.4% of men and 30.6% of women). The difficulty in handling (prickly hairs) was rated as the third constraint by women (24.7%) while unavailability of seed in agrovets was rated third by men (16%). However, farmers were specific about the most important arthropod pests. Spider smites were reported to have been observed on own farms by 50.8% of men and 63.1% of women, while sorghum shoot flies had been observed by 58.1% of men and 67.9% of women. The seriousness of spider mites was mostly perceived by both men and women as moderate problem (63.4% and 44.7%, respectively) to severe problem (21% and 33.9%, respectively). Sorghum shoot flies were also regarded by both men and women as a moderate problem (58.8% and 52.2%, respectively) to severe problem (15.8% and 22.4%, respectively). Further, more farmers (55.6% men and 63.2% women) had observed spider mites on other farms. On the other hand, more men (50.3%) had not observed sorghum shoot flies on other farms, while more women (53.1%) have observed.

Table 2. Perception on brachiaria production challenges disaggregated by gender.

Variable	Response/Rating (%)	Homabay ($n = 38$)		Mbita ($n = 38$)		Bondo ($n = 67$)		Siaya ($n = 54$)		Tanzania ($n = 26$)		Mean ($n = 45$)	
		Male	Female	Male	Female	Male	Female	Male	Female	Male	Female	Male	Female
Access to planting materials	Do not know where to get seeds	5	0	11.8	23.8	22.6	13.9	0	0	6.3	0	9.1	7.5
	Planting materials are expensive	15	0	0	0	0	2.8	20	5.9	0	0	7	1.7
	Unavailability of seeds in Agrovets	30	33.3	0	0	0	2.8	25	23.5	25	0	16	11.9
Planting	Time consuming	0	5.6	0	0	3.2	0	10	8.8	6.3	0	3.9	2.9
	Do not know planting procedure	0	0	5.9	0	3.2	8.3	0	0	0	0	1.8	1.7
	Poor germination of seeds	0	5.6	5.9	0	0	0	5	5.9	37.5	20	9.7	6.3
Crop management	Difficulty in controlling weeds	15	11.1	11.8	4.8	3.2	0	10	23.5	0	0	8	7.9
	Difficulty in handling (it pricks)	5	5.6	29.4	57.1	0	2.8	20	38.2	12.5	20	13.4	24.7
Arthropod pest attacks	Yes	50	66.7	75	60	68.4	77.8	27.8	31.3	25	80	49.2	63.1
	No	50	33.3	25	40	31.6	22.2	72.2	68.8	75	20	50.8	36.9
Spider mites on own farm	Yes	58	66.7	75	60	68.4	77.8	27.8	31.3	25	80	50.8	63.1
	No	42	33.3	25	40	31.6	22.2	72.2	68.8	75	20	49.2	36.9
Seriousness of mites on own farms	No problem	18.2	11.1	12.5	18.8	5.3	4.5	12.5	46.7	0	0	9.7	16.2
	Moderate problem	45.5	22.2	37.5	37.5	78.9	54.5	75	46.7	80	62.5	63.4	44.7
	Severe problem	18.2	66.7	43.8	31.3	10.5	27.3	12.5	6.7	20	37.5	21	33.9
	Very severe problem	9.1	0	6.3	12.5	5.3	13.6	0	0	0	0	4.1	5.2
Spider mites seen on other farms	Yes	46.7	57.1	68.8	52.9	94.7	91.7	27.8	34.5	40	80	55.6	63.2
	No	53.3	42.9	31.3	47.1	5.3	8.3	72.2	65.5	60	20	44.4	36.8
Seen shoot flies on own farm	Yes	57.1	66.7	81.3	61.9	93.3	100	5.6	21.9	53.3	88.9	58.1	67.9
	No	42.9	33.3	18.8	38.1	6.7	0	94.4	78.1	46.7	11.1	41.9	32.1
Seriousness of shoot flies	No problem	20	7.7	13.3	20	6.7	0	60	46.2	0	0	20	14.8
	Moderate problem	60	46.2	33.3	33.3	73.3	60	40	53.8	87.5	77.8	58.8	54.2
	Severe problem	20	30.8	26.7	40	20	30	0	0	12.5	11.1	15.8	22.4
	Very severe problem	0	15.4	26.7	6.7	0	10	0	0	0	11.1	5.3	8.6
Seen shoot flies on other farms	Yes	50	62.5	76.5	55.6	93.3	100	10	3.1	18.8	44.4	49.7	53.1
	No	50	37.5	23.5	44.4	6.7	0	90	93.8	81.3	55.6	50.3	46.3
Disease infestation	Yes	26.7	26.7	64.7	28.6	90.9	81.3	21.1	16.7	18.8	0	44.4	30.6
	No	73.3	73.3	35.3	71.4	9.1	18.8	78.9	83.3	81.3	100	55.6	69.4

3.4. Farmer Evaluation and Selection of Brachiaria Genotypes

Farmers assessed the six brachiaria genotypes based on the following criteria; leaf hairlines, leaf size, leaf softness, number of shoot tillers, plant spread, plant height, seed setting, resistance to the spider mites, and visual estimation of biomass yield. Mulato II ranked the highest in hairlines, leaf softness, and tillers, while Xaraes had the highest numbers for plant height, resistance to spider mites, and biomass yield (Figure 2). Sorghum shoot flies attack the crop at the seedling stage, especially when seed is used as a propagation material. Due to unavailability seeds, we used root splits in our study, further, the crop was evaluated at maturity. Therefore, farmers could not evaluate the materials based on resistance to Sorghum shoot flies. Figure 3 presents the farmer selection of different brachiaria genotypes. Generally, the majority (41.2%) of the farmers preferred Xaraes, followed by Mulato II (25.6%) and Piata (20.4%).

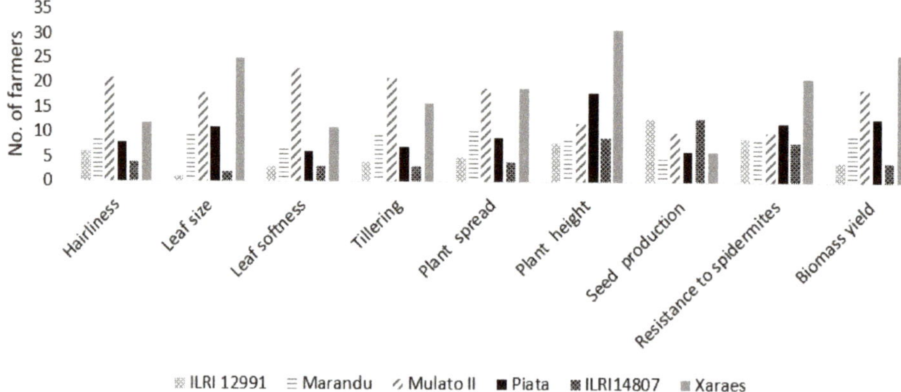

Figure 2. Average number of farmers recorded for the highest score 4 (excellence) for different traits of brachiaria.

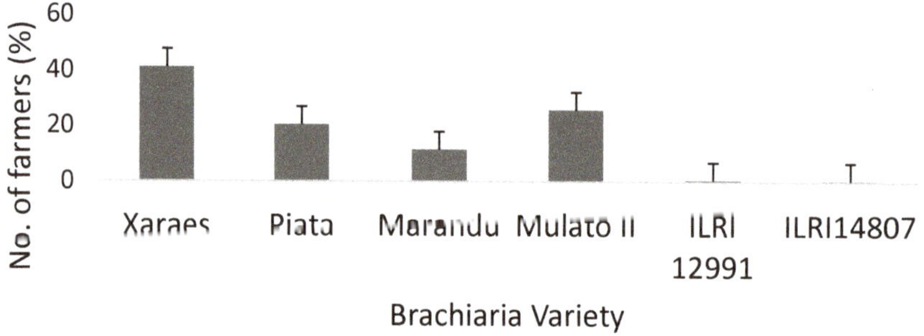

Figure 3. Means for number of farmers (%) for each brachiaria genotype across all the districts. Bars represent standard error of the means.

4. Discussion

The study assessed farmers' experiences and perceptions of brachiaria, a companion crop in a climate-smart PPT. To date, *Brachiaria brizantha* cv. Mulato II is the only variety planted by PPT farmer; therefore, farmers' experience with brachiaria in PPT, as assessed in this study, is based on this variety. However, some farmers may have planted a different variety, but in pure crop stands. The respondents comprised a slightly higher number of females than males. This shows that women are significant and crucial in agricultural development in the region. Studies have shown that, even though most of the African cultures discriminate against women, limiting their land and property rights, they still account for nearly half of the smallholder farmers [30]. Respondents in the study fall in different age categories with the majority being adults between 41–50, while the lowest comprises of the youth between 20–30 years. There is an emerging debate regarding the declining interest of Africa's young people in agriculture [31]; this trend is evident in the current study. Youth can play a key role in agriculture. Unlike the older people, they have greater energy and education and are better equipped to handle modern agricultural technologies and entrepreneurship and they can reverse the ageing trend of the African farming population. It is perhaps worth mentioning that there is a greater interest in sustainability, e.g., of PPT, by those in lower age groups and that this might also improve gender parity in agriculture. One of the major challenges that the young prospective farmers experience is a lack of access to assets and resources that would increase their productivity, such as land, farming inputs, and tools [32]. However, PPT can solve some resource related problems, as all companion crops are perennial and self-saved cereal seed performs better in PPT than do most commercial hybrids [32].

The study provides evidence that brachiaria is an important source of fodder besides its prominent use in pest management strategy (Figure 1). It is used as "pull" component for cereal pests in the climate adapted push-pull technology (PPT); a habitat management strategy that was initially developed to manage the lepidopterous stemborers [7]. In a recent farmer perception study, farmers rated the climate-adapted push-pull as being superior in reducing fall armyworm damage on maize [5]. There is shortage of forages in quantity and quality in sub-Saharan Africa, especially during the dry seasons [33]; therefore, the study validates the value of brachiaria as an ideal forage in the region. Other uses of brachiaria, as mentioned by the farmers, include soil conservation by the prevention of soil erosion, sale, and exchange for milk (Figure 1). Brachiaria grasses are well known for improving soil aggregation thus increasing the resistance to soil degradation and erosion [34]. However, farmers listed several production challenges mainly being unavailability of brachiaria seeds in local retail agents, followed by weed and pest attacks and poor seed germination among others (Table 2). The unavailability of seed is mainly caused by high import costs, cumbersome seed registration processes, relatively undeveloped forage seed market, physical constraints, like drought, low germination rates, and the perceived high opportunity cost of growing the seed in Africa. Furthermore, local brachiaria seed production is underdeveloped, partly due reluctance by the private sector citing unorganized and dispersed demand for seed [35].

Arthropod pests are among the major causes of chronic food insecurity witnessed in the region and are expected to worsen with increasing hot and dry conditions associated with climate change [36]. Strategies to minimize such constrains are crucial in the intensification of smallholder farming systems towards achieving food security in the region. Farmers cited spider mites and sorghum shoot flies as main production challenges and as the main pests of *Brachiaria brizantha* cv. Mulato II. They mostly rated both pests as moderate to severe problem. Susceptibility of Mulato II to spider mites has been reported in previous studies [26]. Common symptoms of pest damage on brachiaria, as described by respondents in a focus group discussion (FGD), include: stunted growth, yellowing of leaves, and wilting of growing tips. Spider mites are tiny and difficult to detect; they are manifested through yellowing of leaves, which the farmers often confuse for mineral deficiency.

Nevertheless, crop improvement based on conventional breeding will continue to be important; many cycles of crossing and backcrossing (pre-breeding) are required to detect and map useful traits [37]. Brachiaria genotypes developed in the Americas and Australia have a higher risk of pest and disease attacks as well as poor adaptability in new environments in Africa. This kind of research involving farmers helps to generate and validate new strategies of integrating crop protection and livestock production which are locally adaptable, and can be introduced widely and applied more rapidly through the discovery of unique traits of companion plants. The introduction of pest-resilient trap plants in push-pull ensures that the technology's full range of opportunities for yield enhancement are exploited in Africa.

When considering the susceptibility of Mulato II to spider mites, there is a need to deploy alternative resistant genotypes, which possess same properties that make them preferred for egg laying by stemborer moths. However, for the successful uptake of such materials, farmers' needs and opinions are key. Farmers evaluated and selected their preferred genotypes from candidate varieties proposed from previous studies, which evaluated drought tolerance, resistance to spider mites, biomass yield, and attractiveness to oviposition by stemborers [25–27]. Xaraes was a highly preferred genotype, followed by Mulato II and Piata (Figure 3). Some of the traits that farmers proposed as a criteria in evaluating brachiaria genotypes are leaf hairlines, leaf size, leaf softness (as a measure of palatability by animals), number of tillers, plant spread, plant height, seed production, resistance to spider mites, and biomass yield. Farmers generally prefer less hairy genotypes for ease in cut and carry, because the hairs are irritating to the skin. They also believe that softer leaves are highly palatable and preferred by the animals, a trait for which they voted Mulato II as superior. However, there is a trade-off between hairlines and leaf softness in Mulato II and this might produce mixed results in farmers' rating of the cultivar.

The results of this study are important to policy makers in sub-Saharan Africa because the sustainable increase in agricultural productivity represents a significant opportunity for addressing the pervasive challenge of low productivity, which results in high poverty levels and under-nourishment. Moreover, climate-smart and resilient agricultural systems that are based on such genetic material are needed to protect and enhance natural resources and ecosystem services in ways that mitigate future climate change [38]. The exploitation of such climate-smart, resilient material help farmers to develop production systems that are compatible with their farming systems, and sound management of available natural resources. The involvement of farmers in scientific developments in agronomy and agroecological practices take into account their other on-farm enterprises, like livestock keeping, and helps them to fully exploit the benefits of production and resource conservation technologies.

5. Conclusions

The study provides evidence that brachiaria is an important source of fodder besides its prominent use in pest management strategy. Its multiple utility facilitates sustainable intensification of smallholder agriculture by facilitating the integration of cereal and livestock fodder production. It is used as "pull" component for cereal pests in the climate adapted push-pull technology (PPT), a habitat management strategy initially developed to manage the lepidopterous stemborers, while it generated quality fodder. The study demonstrates that these Brachiaria genotypes could be of value in the improvement of cereal livestock-based livestock productivity in sub-Saharan Africa in the current scenarios of increasing aridification and attacks by invasive pests, such as spider mite (*Oligonychus trichardti*). It also demonstrates that farmers' skills and knowledge can complement scientific research, and that their contribution through participatory approach is key in validating the potential of such genetic materials.

Author Contributions: Conceptualization, D.C., C.A.M., J.A.P. and Z.R.K.; Formal analysis, J.O.P.; Funding acquisition, C.A.M., J.A.P. and Z.R.K.; Investigation, D.C. and J.O.P.; Methodology, D.C., J.O.P., J.A.P. and Z.R.K.; Project administration, C.A.M. and Z.R.K.; Resources, C.A.M., J.O.P., J.A.P. and Z.R.K.; Supervision, C.A.M. and Z.R.K.; Validation, C.A.M.; Writing—original draft, D.C.; Writing—review & editing, D.C., J.O.P. and J.A.P. All authors have read and agreed to the published version of the manuscript.

Funding: We gratefully acknowledge the financial support for this research by the following organizations and agencies: European Union; Biovision foundation; UK's Department for International Development (DFID); Swedish International Development Cooperation Agency (Sida); the Swiss Agency for Development and Cooperation (SDC); Norwegian Agency for Development Cooperation (NORAD) and the Ethiopian and Kenyan Governments. The views herein do not necessarily reflect the official opinion of the donors.

Acknowledgments: We further thank the International Centre for Tropical Agriculture (CIAT) and International Livestock Research Institute (ILRI) for provision of brachiaria genotypes studied. The authors also acknowledge contributions by the farmers, *icipe* field staff, Kenyan Ministry of Agriculture extension staff and Tanzania Agricultural Research Institute staff.

Conflicts of Interest: The authors declare no conflict of interest.

References

1. Miles, J.W.; Do Valle, C.B.; Rao, I.M.; Euclides, V.P.B. Brachiariagrasses. In *Warm-Season (C₄) Grasses*; Moser, L.E., Burson, B.L., Sollenberger, L.E., Eds.; Agronomy Monographs; American Society of Agronomy, Crop Science Society of America, Soil Science Society of America: Madison, WI, USA, 2004; Volume 45, pp. 745–783.
2. Jank, L.; Alves, G. The value of improved pastures to Brazilian beef production. *Crop Pasture Sci.* **2014**, *65*, 1132–1137. [CrossRef]
3. Boonman, G. *East Africa's Grasses and Fodders: Their Ecology and Husbandry*; Springer: Heidelberg, Germany, 1993.
4. Maass, B.L.; Midega, C.A.O.; Mutimura, M.; Rahetlah, V.B.; Salgado, P.; Kabirizi, J.M.; Khan, Z.R.; Ghimire, S.R.; Rao, I.M. Homecoming of *Brachiaria*: Improved hybrids prove useful for African animal agriculture. *East Afr. Agric. For. J.* **2015**, *81*, 71–78. [CrossRef]
5. Midega, C.A.O.; Pittchar, J.O.; Pickett, J.A.; Hailu, G.W.; Khan, Z.R. A climate-adapted push-pull system effectively controls fall armyworm, *Spodoptera frugiperda* (J E Smith), in maize in East Africa. *Crop Prot.* **2018**, *105*, 10–15. [CrossRef]
6. Khan, Z.R.; James, D.G.; Midega, C.A.O.; Pickett, J.A. Chemical ecology and conservation biological control. *Biol. Control.* **2008**, *45*, 210–224. [CrossRef]
7. Khan, Z.; Midega, C.; Pittchar, J.; Pickett, J.; Bruce, T. Push–pull technology: A conservation agriculture approach for integrated management of insect pests, weeds and soil health in Africa. *Int. J. Agric. Sustain.* **2011**, *9*, 162–170. [CrossRef]
8. Khan, Z.; Midega, C.A.O.; Hooper, A.; Pickett, J. Push–Pull: Chemical Ecology-Based Integrated Pest Management Technology. *J. Chem. Ecol.* **2016**, *42*, 689–697. [CrossRef]
9. Khan, Z.R.; Midega, C.A.O.; Pittchar, J.; Bruce, T.J.A.; Pickett, J.A. 'Push–pull' revisited: The process of successful deployment of a chemical ecology-based pest management tool. In *Biodiversity and Insect Pests: Key Issues for Sustainable Management*; Gurr, G.M., Wratten, S.D., Snyder, W.E., Read, D.M.Y., Eds.; John Wiley & Sons Ltd.: Hoboken, NJ, USA, 2012; pp. 259–275.
10. Pickett, J.A.; Woodcock, C.M.; Midega, C.A.O.; Khan, Z.R. Push–pull farming systems. *Curr. Opin. Biotechnol.* **2014**, *26*, 125–132. [CrossRef]
11. Khan, Z.R.; Pickett, J.A.; Wadhams, L.; Muyekho, F. Habitat management strategies for the control of cereal stemborers and striga in maize in Kenya. *Int. J. Trop. Insect Sci.* **2001**, *21*, 375–380. [CrossRef]
12. Khan, Z.R.; Midega, C.A.O.; Bruce, T.J.A.; Hooper, A.M.; Pickett, J.A. Exploiting phytochemicals for developing a 'push–pull' crop protection strategy for cereal farmers in Africa. *J. Exp. Bot.* **2010**, *61*, 4185–4196. [CrossRef]
13. Pickett, J.A.; Khan, Z.R. Plant volatile-mediated signaling and its application in agriculture: Successes and challenges. *New Phytol.* **2016**, *212*, 856–870. [CrossRef]
14. Khan, Z.; Hassanali, A.; Pickett, J. Managing polycropping to enhance soil system productivity: A case study from Africa. In *Biological Approaches to Sustainable Soil Systems*; Ball, A.S., Thies, J., Sanginga, N., Sanchez, P., Pretty, J., Palm, C., Laing, M., Herren, H., Fernandes, E., Uphoff, N., Eds.; CRC Press: Boca Raton, FL, USA, 2006; pp. 575–586.
15. Midega, C.A.O.; Wasonga, C.J.; Hooper, A.M.; Pickett, J.A.; Khan, Z.R. Drought-tolerant *Desmodium* species effectively suppress parasitic striga weed and improve cereal grain yields in western Kenya. *Crop Prot.* **2017**, *98*, 94–101. [CrossRef] [PubMed]
16. Mutimura, M.; Ebong, C.; Rao, I.M.; Nsahlai, I.V. Effects of supplementation of *Brachiaria brizantha* cultivar Piatá and Napier grass with *Desmodium distortum* on feed intake, digesta kinetics and milk production by crossbred dairy cows. *Animal Nutr.* **2018**, *4*, 222–227. [CrossRef] [PubMed]
17. Subbarao, G.V.; Nakahara, K.; Hurtado, M.P.; Ono, H.; Moreta, D.E.; Salcedo, A.F.; Yoshihashi, A.T.; Ishikawa, T.; Ishitani, M.; Ohnishi-Kameyama, M.; et al. Evidence for biological nitrification inhibition in *Brachiaria* pastures. *Proc. Natl. Acad. Sci. USA* **2009**, *106*, 17302–17307. [CrossRef] [PubMed]

Article

Response of Yellow Lupine to the Proximity of Other Plants and Unplanted Path in Strip Intercropping

Lech Gałęzewski *, Iwona Jaskulska, Edward Wilczewski and Anna Wenda-Piesik

Department of Agronomy, Faculty of Agriculture and Biotechnology, University of Science and Technology, 7 prof. S. Kaliskiego St., 85-796 Bydgoszcz, Poland; jaskulska@utp.edu.pl (I.J.); edward@utp.edu.pl (E.W.); apiesik@utp.edu.pl (A.W.-P.)
* Correspondence: lechgalezewski@op.pl; Tel.: +48-523-749-499

Received: 9 June 2020; Accepted: 6 July 2020; Published: 10 July 2020

Abstract: Taking into account the climatic conditions of central Europe, yellow lupine is often considered as an alternative to soybean, which has significantly higher thermal requirements. Attempts to intercrop yellow lupine with cereals have often resulted in failure. In combined production, the relative amount of lupine has proven to be considerably smaller given the sowing mix proportions and its yield potential in pure stand. Low yield is attributed to lupine's low competitive potential, therefore strip intercropping presents a viable alternative. The main goal of the experiment was to determine the response of yellow lupine to the neighboring presence of wheat, triticale, barley, and pea, as well as to estimate the production effects of lupine in strip intercropping. Field trials were carried out in Poland (53°13′ N; 17°51′ E) in the years 2008–2010. The experimental factor consisted of row layout: a four-row separation between lupine and the neighboring species. The proximity of cereals and peas proved to be most unfavorable to yellow lupine. It was determined that yellow lupine was most intolerant of barley and least affected by the proximity of peas. Depending on the neighboring species, adverse effects extended up to the third row of lupine's canopy. A beneficial alternative for the production effect involves an introduction of a path separating the lupine strip from the tested species.

Keywords: proximity effect; border effect; neighbor effect; strip intercropping; legume; cereals

---

1. Introduction

Soybean seed is the world's primary source of plant protein. In temperate climate conditions soybean cultivation remains relatively inefficient [1]. Therefore, yellow lupine seeds [2–4] have become a promising alternative protein source. Cultivation can be carried out by pure sowing or by intercropping with other spring crop species.

Due to intercropping of lupine with other plants, the seed yield decreases; however, the protein yield is noticeably higher [5]. For coexisting species, we distinguish different intercropping systems depending on the time of sowing and spatial placement [6]. Cereals (wheat, barley, and oats) and legumes (yellow lupine, narrow leaf lupine, and peas) are grown mainly in mixed intercropping (MI), i.e., they are planted in the same rows.

As a consequence of MI, the optimization of fertilization techniques is considerably limited, and the implementation of herbicide control is not possible. Corn, along with other species, is grown by utilizing strip intercropping (SI), i.e., alternating strips of various species [7–10]. If strips of a single species are wide enough and adjusted to the technical capabilities of cultivating tools, it also becomes possible to optimize the cultivation practice for individual species.

In integrated conditions, particularly in organic production systems, the cultivation of cereals with legumes in mixtures is considered to be a suitable source of concentrated feed [11,12]. Intercropping also serves ecological functions: it increases biodiversity, positively affects the soil condition, and suppresses weed infestation [13]. Therefore, the inability to optimize cultivation practice and difficulties regarding herbicide control should not constitute an obstacle or become a deterrent in MI implementation. Because of enhanced utilization of habitat capacity, MI crops are generally more stable in subsequent years as compared to pure crop yields of species used in MI [14–19]. Unfortunately, the co-occurrence of individual species may also contribute to unfavorable effects. The adverse effects vary considerably, and they are strictly dependent on weather conditions. As a consequence, varying qualities of yields are obtained in subsequent growing seasons regardless of the fact that the same agrotechnical assumptions are being implemented [20], which, in turn, leads to difficulty in balancing feed resources [21].

In the scientific literature, the subject of SI primarily relates to the cultivation of soybeans and corn [22]. Available resources pertaining to SI of yellow lupine with other plant species are rather limited [23]. It is known, however, that the yield of yellow lupine seeds in MI with oats is largely dependent on environmental factors. MI in low moisture soil conditions leads to competition between lupine and oats, and it shows to be asymmetric to the detriment of lupine. Consequently, MI results in a considerably smaller yield of yellow lupine [24,25]. Since the interaction between species occurs exclusively at the strips' border, it has been ascertained that SI cultivation of yellow lupine with spring cereals can be justified. In SI, yellow lupine adversely responds to close proximity with oats and triticale [23]. However, the response of yellow lupine in the proximity of other plant species (potential components for SI) remains undetermined. While taking into consideration the asymmetry of competition between various species, row separation with a technological path presents a viable option. Therein lies the advantage of SI over MI. Separating the species tends to diminish the competition effect and utilize the positive phenomenon of the border effect: namely, an increase in the yield of plants cultivated adjacent to an area devoid of vegetation [26,27].

The aim of our study was to determine the proximity effect of spring wheat, triticale, barley, and peas on yellow lupine cultivation and to estimate its yield in strip intercropping with the abovementioned plant species.

2. Materials and Methods

2.1. Experiment Site

The field trial was carried out between 2008 and 2010. The experiment was conducted at the Research Station of the Faculty of Agriculture and Biotechnology in Mochełek (53°13′ N; 17°51′ E) (Figure 1). The results presented in this manuscript are part of previously published studies related to the proximity effect (PE) on other species [28–31]. Accordingly, the methodology presented in this experiment coincides with the cited studies.

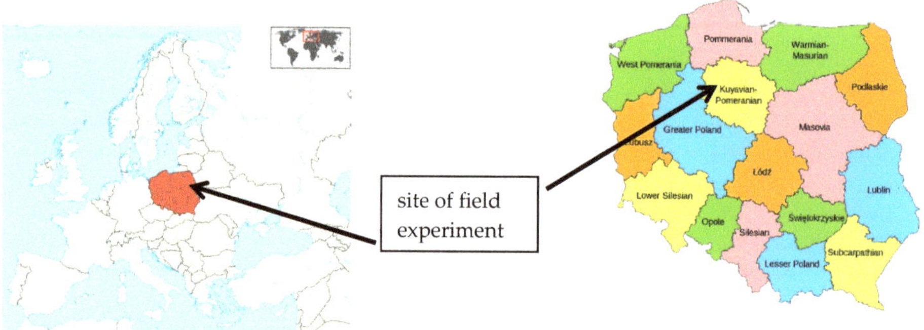

Figure 1. Site of field experiment at Mochełek, Kuyavian-Pomeranian voivodeship, Poland [32,33].

The experiment was conducted on loam sand texture luvisol soil (LV) [34]; the pre-crop was winter oil seed rape. Depending on the research year, the Corg content was 6.2–6.6 g·kg^{-1} d.m. of soil, and the content of absorbable forms P and K was 63–69 and 94–172 mg·kg^{-1} respectively, soil pH (1M KCl) was between 5.2–6.6.

During the growing season, the temperature amplitude was similar for all three years of the conducted research (Figure 2). In 2009, however, April and the first two decades of June were characterized by warmer temperatures as compared to the rest of the year. The year 2010 was marked by a much warmer July. Distribution of rainfall also varied significantly. Modest precipitation was observed from the third decade of April to the second decade of June of 2008. During this time period, in any of the decades, the rainfall did not exceed 10 mm. In 2010, rainfall not exceeding 10 mm per decade was reported between the first decade of June and the second decade of July.

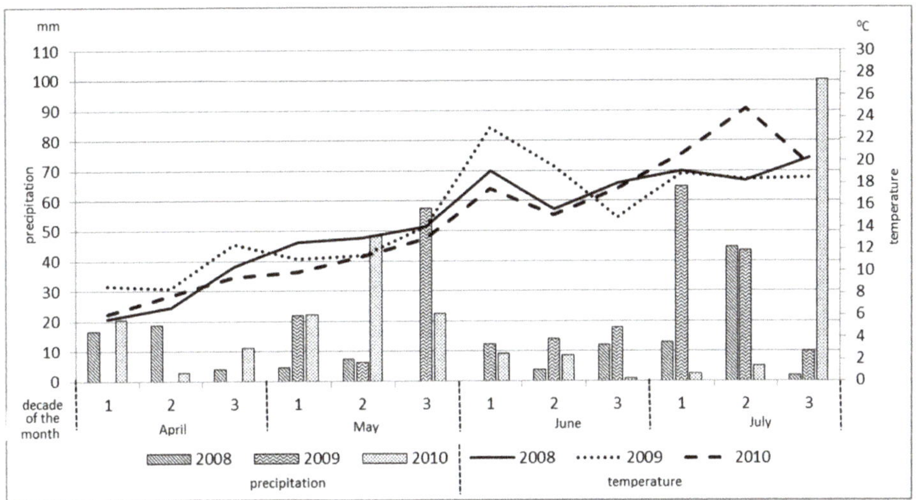

Figure 2. Precipitation and air temperature (2008–2010) at the site of the field experiment.

2.2. Experiment Design

The source data come from a multiple 3-year field experiment. The layout of the experiment is demonstrated in Figure 3. The plot was 150 cm wide and consisted of 12 rows of plants separated by 12.5 cm. Figure 3A represents one of four replications (randomized complete blocks) with all the neighboring species of yellow lupine and their paths. The experimental treatment consisted of yellow lupine's row layout (Figure 3B), four rows of separation termed PE (proximity effect) in relation to the neighboring species (wheat, triticale, barley, and oat) or separated from an unplanted path referred to as BE (border effect) (Figure 3C). The first adjacent row was located 12.5 cm from the first row of the neighboring species/path. The experimental plot consisted of successive plant rows each measuring four meters long. The mean result of each treatment of adjacent plants (from right and left sides of the plot) was considered as a single replication. Based on the results of previous studies [23], the fourth plant row was no longer subjected to the influence of neighboring plants, representing the internal canopy (control). The orientation of the plots' longer side was north-south.

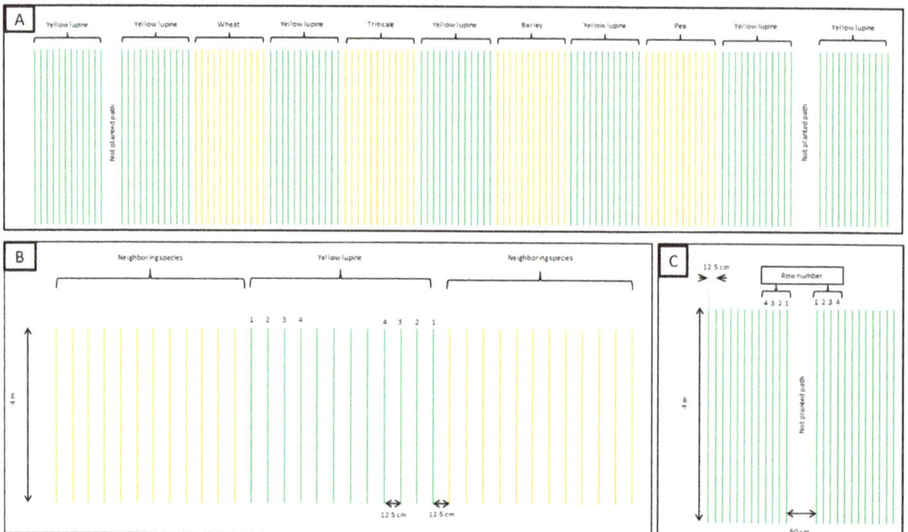

Figure 3. Experiment design: single block (**A**), PE single plot (**B**), and border effect (BE) single plot design (**C**).

2.3. Elements of Agrotechnical Practices

Each plant species was sown simultaneously between 25 March and 5 April. In order to ensure even spacing between each plant, the cereal seeds were precisely placed on a seeding belt made of blotting paper. Plant density was 45 pcs·m^{-1} (360 pcs·m^{-2}). The seeding strips were placed in the soil at a depth of 4 cm. Seeds of lupine and peas were sown manually; planting density was 10 pcs·m^{-1} (80 pcs·m^{-2}).

The following cultivars were planted: yellow lupine 'Lidar', spring wheat 'Bombona', spring triticale 'Doublet', spring barley 'Antek', and pea 'Ramrod'.

Macro-nutrients were applied during the spring months: 30 kg P·ha^{-1}, 66 kg K·ha^{-1} and 34 kg N·ha^{-1}. In the phenological phase BBCH 22–25 (tillering stage), N fertilization (34 kg N·ha^{-1} dose) was used with cereals only. Herbicide active substance-linuron (Alfalon 450SC), at a dose of 1 dm^3·ha^{-1} was applied to each crop.

2.4. Samples and Measurements

Harvest sampling from each row was conducted manually. The measurements of yellow lupine plants included:

- Plant density (number of plants with no less than one pod with seeds in particular rows were considered)
- Pod density (number of pods containing no less than one seed in particular rows were considered)
- Pods per plant (from calculation: pod density/plant density)
- Seed per pod (harvested pods were threshed and the number of seeds was determined)
- Thousand-seed weight (TSW)
- Straw weight (biomass without pods)
- Biomass (straw weight + seed weight)
- Seed weight (g per row)

Weight was recalculated for 1 m of the row.

2.5. Data Analysis

Single year data concerning all characteristics of yellow lupine in strip intercropping were calculated using one-way ANOVA in a four reps (block) model. The three-year synthesis of variance, based on statistic F (Fisher) in a mixed model, tested the null hypotheses regarding year as random effect and treatments (yellow lupine's row neighboring to one of four species) as fixed effect (Table 1). The post-hoc calculation according to HSD Tukey's test ($p = 0.05$) was used for the separation of means of yellow lupine traits. For data verification, the R core team software package was used.

Table 1. Significance of factor and significance of interaction factor and years in ANOVA.

Characteristic of Yellow Lupine	Variation Source	Species				Path
		Wheat	Triticale	Barley	Pea	
Plant density	Factor	**	**	**	**	**
	factor × year	-	-	**	-	-
Pod density	Factor	**	**	**	**	**
	factor × year	*	*	*	-	-
Pods per plant	Factor	**	**	-	-	**
	factor × year	-	-	*	-	**
Seeds per pod	Factor	*	**	-	-	-
	factor × year	*	-	-	-	-
TSW	Factor	**	*	**	-	**
	factor × year	-	-	-	-	*
Straw weight	Factor	**	**	**	**	**
	factor × year	**	-	-	*	*
Biomass	Factor	**	**	**	**	**
	factor × year	**	**	**	**	**
Seed weight	Factor	**	**	**	**	**
	factor × year	**	**	**	**	**

* significant $p < 0.05$; ** significant $p < 0.01$; - not significant.

Index of the proximity effect (IPE) was based on the results acquired from three rows closest to the neighboring species; IPE reflects the quotient of trait values for the given order and the fourth order.

$$PE = \frac{R_{(1,2,3)}}{R_{(4)}} \tag{1}$$

where $R_{(1,2,3)}$ is the seed weight of plants from 1st or 2nd or 3th row; and $R_{(4)}$ is the seed weight of plants from the 4th row.

IPE = 1 implies neutrality of the tested species. IPE < 1 indicates a negative impact of the neighboring species on yellow lupine. IPE > 1 indicates positive influence of the neighboring species on yellow lupine. Index of the border effect (IBE) was calculated as well. In this instance, the yellow lupine plants were adjacent to a vegetation-free area and separated by a technological path or a path dividing the plots. The interpretations of the IBE and IPE values are the same.

The proposed predictive analysis is to adopt the results of the yield from this study to the practical utilization of yellow lupine in SI with various species. As the sowing is practiced by a 3-m-wide seed driller, we applied, in reference to yield estimation for each linear meter, 3-m-wide strips (24 rows), with a row spacing of 12.5 cm. Estimated yield (Figures 4 and 5) was calculated based on the following formulas:

$$Y_{no\ proximity} = 24 \times r_4 \tag{2}$$

$$Y_{one\ side\ proximity} = r_1 + r_2 + r_3 + 21 \times r_4 \tag{3}$$

$$Y_{two\ side\ proximity} = 2 \times r_1 + 2 \times r_2 + 2 \times r_3 + 18 \times r_4 \tag{4}$$

where r_{1-4} represents the yield in the next row from the neighboring species.

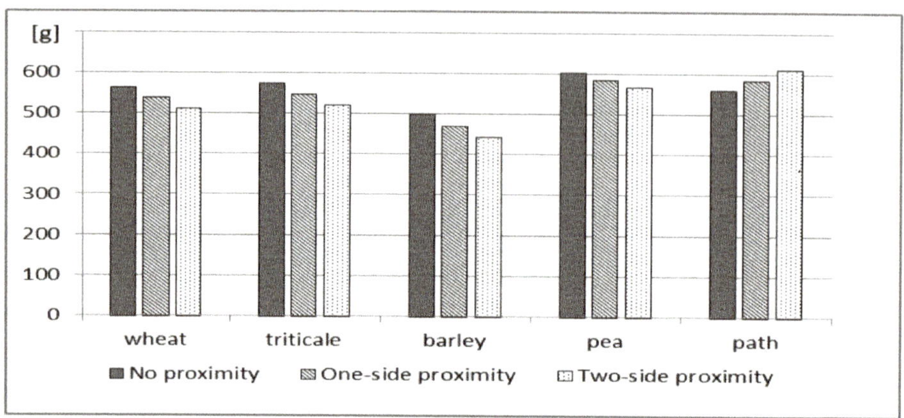

Figure 4. Estimated yellow lupine yield (g) for each linear meter of 3-m-wide strips, depending on the type of proximity.

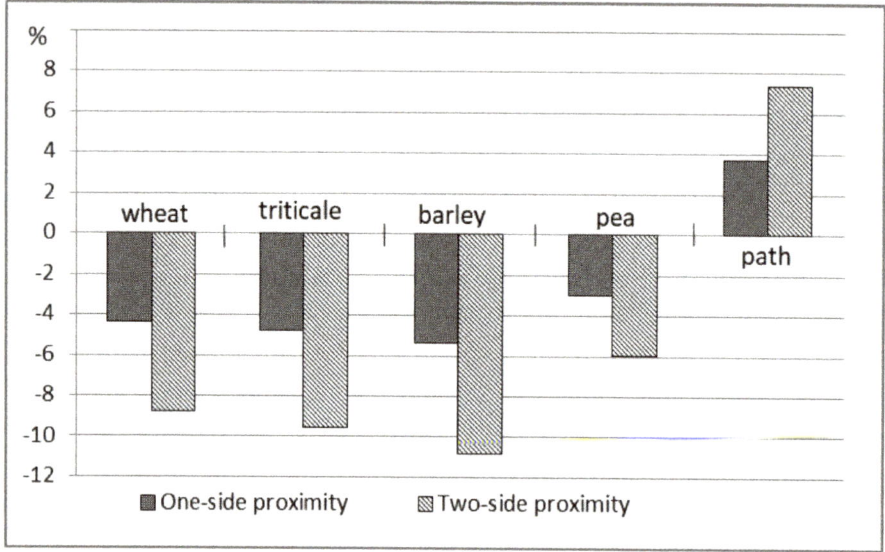

Figure 5. Estimated yellow lupine yield difference (%), depending on neighboring species/path.

The total yield and yield structure of SI for two species in an area of one hectare were also estimated (Figure 6). These crops were estimated considering the immediate vicinity of strips, and strips separated by a 50-cm-wide path. For estimates, 17 rows, each 3-m-wide were adopted for both species (34 rows in total). The above setup resulted in arable fields of 102 × 98 m for SI without paths and 114.75 × 87.1 m for SI with paths. Estimated cereal yields were based on the results from the same experiment that was already published in other articles [28–31].

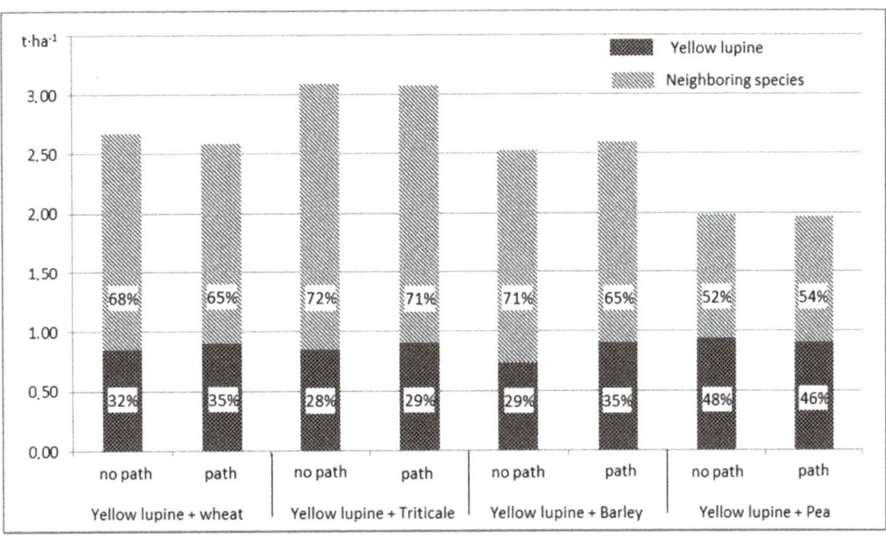

Figure 6. Estimated strip intercropping (SI) yield, and yield structure, depending on neighboring species/path.

3. Results

Straw, seed, and plant weight of yellow lupine was significantly affected by the PE of all tested species (Table 1). Regarding the weight of plant and lupine seeds, the PE varied throughout the years, and its effect was manifested in all of the species. In the case of barley, the PE affected the density of plants and pods, as well as the lupine's TSW. The PE of peas, in addition to having an effect on the previously mentioned weight of straw, seed, and the lupine plants' weight, also affected the density of lupine's plants and pods.

As evidenced by the IPE index values below one, wheat strip proximity proved to have an adverse effect on yellow lupine plants (Table 2). The density of lupine plants increased significantly in rows furthest from wheat (up to the fourth row). In rows directly adjacent to the wheat strip, density of yellow lupine plants was 22.6% lower as compared to the fourth row. The density of pods also increased in rows that were located furthest from wheat, but the statistically confirmed effect was obtained for the first and second rows. In the immediate vicinity of wheat, i.e., in the first row, the pod density was 42.2% lower than in the fourth row. A negative PE was found in the first row only; it influenced the number of pods per plant, the amount of seeds in the pod, and TSW. The straw and lupine plants' weight also increased proportionally to the distance from wheat; this effect was confirmed in the first and second rows. The IPE indicates that the negative effect of wheat on lupine was most evident in lupine's weight (IPE = 0.55) and pod density (IPE = 0.58).

Similarly to wheat, the PE of triticale also proved to be unfavorable to yellow lupine. In rows 1, 2, and 3, for each of the characteristics, IPE values were less than one (Table 3). For triticale, the IPE was comparable to the values obtained for wheat. In the successive rows, a tendency for values to increase manifested itself and was evident for all presented characteristics. With the exception of straw weight, statistically confirmed unfavorable PE was limited to the second row and its effect was present in all the characteristics. In the case of straw weight, an adverse effect of triticale's neighboring presence was confirmed only in the first row. An unfavorable PE was least evident when taking into account the thousand-seed weight, which in the first row was 4.0% lower as compared to the fourth row. The proximity of the triticale strip resulted in the reduction of plant weight: the weight of the first row was 44.1% less than the weight of the fourth row.

Table 2. Response of yellow lupine plants to the proximity of spring wheat.

Characteristic of Yellow Lupine	Unit	Subsequent Plot Row			
		1	**2**	**3**	**4**
Plant density	(plant·m^{-1})	4.10 [d]	4.55 [c]	4.90 [b]	5.30 [a]
	IPE *	0.77	0.86	0.92	-
Pod density	(pod·m^{-1})	27.0 [c]	34.9 [b]	41.7 [ab]	46.7 [a]
	IPE	0.58	0.75	0.89	-
Pod per plant	Pod	6.5 [b]	7.6 [ab]	8.4 [a]	8.7 [a]
	IPE	0.75	0.87	0.97	-
Seeds per pod	Seed	3.3 [b]	3.6 [ab]	4.1 [a]	4.0 [a]
	IPE	0.83	0.9	1.03	-
TSW	G	115.7 [b]	118.9 [ab]	119.0 [a]	121.7 [a]
	IPE	0.95	0.98	0.98	-
Straw weight	(g·m^{-1})	17.8 [c]	23.1 [b]	25.8 [ab]	27.9 [a]
	IPE	0.64	0.83	0.92	-
Biomass	(g·m^{-1})	28.3 [c]	37.9 [b]	46.3 [ab]	51.4 [a]
	IPE	0.55	0.74	0.9	-

The same letter in a given row indicates the lack of significant differences between means; * proximity effect index, see Section 2.5.

Table 3. Response of yellow lupine plants to the proximity of spring triticale.

Characteristic of Yellow Lupine	Unit	Subsequent Plot Row			
		1	**2**	**3**	**4**
Plant density	(plant·m^{-1})	4.00 [c]	4.30 [bc]	4.60 [ab]	5.00 [a]
	IPE *	0.80	0.87	0.93	-
Pod density	(pod·m^{-1})	28.9 [b]	31.6 [b]	43.7 [a]	47.5 [a]
	IPE	0.61	0.67	0.92	-
Pods per plant	Pod	7.0 [b]	7.1 [b]	9.2 [a]	9.3 [a]
	IPE	0.75	0.76	0.99	-
Seeds per pod	Seed	3.2 [b]	3.3 [b]	3.8 [ab]	4.0 [a]
	IPE	0.80	0.83	0.95	-
TSW	G	117.0 [b]	118.2 [ab]	120.3 [ab]	121.9 [a]
	IPE	0.96	0.97	0.99	-
Straw weight	(g·m^{-1})	18.5 [b]	23.4 [ab]	26.7 [a]	29.0 [a]
	IPE	0.64	0.81	0.92	-
Biomass	(g·m^{-1})	29.6 [b]	36.1 [b]	47.3 [a]	53.0 [a]
	IPE	0.56	0.68	0.89	-

The same letter in a given row indicates the lack of significant differences between means; * proximity effect index, see Section 2.5.

The proximity of barley proved to have an adverse effect on yellow lupine (Table 4). For each of the presented characteristics, the IPE values were less than one. Furthermore, they were considerably different from the values obtained from the previously described cereal species. With the exception of the number of pods, trends of increasing values in subsequent rows were noted for all presented characteristics. Regarding the number of pods, the PE of barley was not statistically confirmed, since it was only 1.1% less in the first row than in the fourth row. Considering the number of seeds in a pod, a negative PE of barley has not been confirmed, although the corresponding difference was much higher (18.4%).

The presence of a barley strip proved to have a negative influence on up to the third row of lupine plants. In comparison to other traits, plant density, pod density and biomass were affected the most.

In the case of the thousand-seed weight and straw weight, the negative effects were evident up to the second row. As compared to other traits, plant biomass has been reduced the most. Between the first and fourth rows, the difference in plants' weight reached 45.3%.

Table 4. Response of yellow lupine plants to the proximity of spring barley.

Characteristic of Yellow Lupine	Unit	Subsequent Plot Row			
		1	2	3	4
Plant density	(plant·m^{-1})	3.30 [c]	4.05 [b]	4.40 [b]	5.05 [a]
	IPE *	0.65	0.80	0.88	-
Pod density	(pod·m^{-1})	27.6 [c]	32.3 [bc]	36.8 [b]	46.6 [a]
	IPE	0.59	0.69	0.79	-
Pods per plant	Pod	8.9 [a]	7.9 [a]	8.2 [a]	9.0 [a]
	IPE	0.99	0.88	0.91	-
Seeds per pod	Seed	3.1 [a]	3.3 [a]	3.6 [a]	3.8 [a]
	IPE	0.82	0.87	0.95	-
TSW	G	112.9 [b]	116.4 [ab]	119.7 [a]	121.3 [a]
	IPE	0.93	0.96	0.99	-
Straw weight	(g·m^{-1})	16.2 [b]	19.0 [b]	21.9 [ab]	25.8 [a]
	IPE	0.63	0.74	0.85	-
Biomass	(g·m^{-1})	25.4 [c]	30.7 [bc]	36.4 [b]	46.6 [a]
	IPE	0.55	0.66	0.78	-

The same letter in a given row indicates the lack of significant differences between means; * proximity effect index, see Section 2.5.

In instances where yellow lupine was grown in the proximity of peas, the IPE for particular traits generated significantly higher values as compared to previously described cereals. This signifies yellow lupine's higher tolerance for neighboring peas as opposed to wheat, triticale or barley. However, IPE values for all considered traits did not exceed one (Table 5). Consequently, it can be inferred that the PE of peas was unfavorable. The influence could not be confirmed in respect to the number of pods per plant, number of seeds in the pod, and TSW. For the remaining characteristics, the negative impact of PE was perceptible only in the first and second rows of lupine plants. As compared to the fourth row, the reduction in plants' weight in the first row was most significant (32.1%).

Table 5. Response of yellow lupine plants to the proximity of pea.

Characteristic of Yellow Lupine	Unit	Subsequent Plot Row			
		1	2	3	4
Plant density	(plant·m^{-1})	4.60 [c]	4.90 [bc]	5.20 [ab]	5.40 [a]
	IPE *	0.84	0.90	0.94	-
Pod density	(pod·m^{-1})	36.6 [c]	39.6 [bc]	45.0 [ab]	47.4 [a]
	IPE	0.77	0.83	0.95	-
Pods per plant	Pod	7.8 [a]	8.0 [a]	8.6 [a]	8.6 [a]
	IPE	0.91	0.93	1.00	-
Seeds per pod	Seed	3.8 [a]	4.0 [a]	4.1 [a]	4.1 [a]
	IPE	0.93	0.98	1.00	-
TSW	G	114.8 [a]	117.5 [a]	121.0 [a]	124.0 [a]
	IPE	0.93	0.95	0.98	-
Straw weight	(g·m^{-1})	21.0 [c]	23.6 [bc]	26.6 [ab]	29.4 [a]
	IPE	0.71	0.80	0.91	-
Biomass	(g·m^{-1})	37.1 [c]	42.2 [bc]	49.4 [ab]	54.6 [a]
	IPE	0.68	0.77	0.91	-

The same letter in a given row indicates the lack of significant differences between means; * proximity effect index, see Section 2.5.

Considering the productive characteristics of yellow lupine (Tables 2–5), strip intercropping with other spring species turned out to be disadvantageous. In most cases, the values of the proximity effect were less than one. A different effect was obtained in the case of lupine plants adjacent to non-grown rows separated by paths (Table 6). Generally, for each feature, an increase in its value was found in the first row (IBE = 1.06 for TSW; 1.59 for biomass) and this positive effect was noticeable up to the third row (IBE = 1.04 for TSW; 1.19 for straw weight). Only for the number of seeds in the pod did the commented tendency not receive statistical confirmation (IBE oscillated 1.05) (Table 6).

Table 6. Response of yellow lupine plants to the border effect.

Characteristic of Yellow Lupine	Unit	Subsequent Plot Row			
		1	2	3	4
Plant density	(plant·m^{-1})	6.00 [a]	5.70 [ab]	5.70 [bc]	5.30 [c]
	IBE *	1.13	1.08	1.08	-
Pod density	(pod·m^{-1})	72.7 [a]	54.5 [b]	47.2 [b]	47.7 [b]
	IBE	1.52	1.14	0.99	-
Pods per plant	Pod	12.8 [a]	11.0 [ab]	9.3 [bc]	8.5 [c]
	IBE	1.51	1.30	1.09	-
Seeds per pod	Seed	4.3 [a]	4.3 [a]	4.1 [a]	3.9 [a]
	IBE	1.12	1.11	1.05	-
TSW	G	128.2 [a]	127.7 [a]	126.0 [a]	121.1 [b]
	IBE	1.06	1.05	1.04	-
Straw weight	(g·m^{-1})	48.1 [a]	38.9 [ab]	36.0 [b]	30.2 [b]
	IBE	1.59	1.29	1.19	-
Biomass	(g·m^{-1})	85.5 [a]	68.1 [b]	61.3 [bc]	53.8 [c]
	IBE	1.59	1.27	1.14	-

The same letter in a given row indicates the lack of significant differences between means; * border effect index, see Section 2.5.

A significant decrease in lupine seed yield in the first and second rows was confirmed in proximity to wheat, triticale, and peas (Table 7). In the case of barley, a negative PE was also confirmed for the third row. The presence of neighboring cereals resulted in a lower yield of yellow lupine seeds in the first row in relation to the fourth row by 53.3% (for triticale) and 55.7% (for barley). A corresponding difference for peas was less significant and reached 36.1%. The border effect resulting from path proximity had a positive effect on seed yield. There was a 58.2% increase in yield in the first row, 23.2% in the second row, and 6.7% in the third row (IBE = 1.58, 1.23, and 1.07, respectively) (Table 7).

Table 7. Yellow lupine seed yield depending on the neighboring species.

Neighboring Species	Unit	Subsequent Plot Row			
		1	2	3	4
Wheat	(g·m^{-1})	10.5 [b]	14.8 [b]	20.5 [a]	23.5 [a]
	IPE *	0.44	0.63	0.87	-
Triticale	(g·m^{-1})	11.2 [b]	12.7 [b]	20.6 [a]	24.0 [a]
	IPE	0.46	0.53	0.86	-
Barley	(g·m^{-1})	9.2 [c]	11.8 [bc]	14.6 [b]	20.8 [a]
	IPE	0.44	0.56	0.70	-
Pea	(g·m^{-1})	16.1 [c]	18.6 [bc]	22.8 [ab]	25.2 [a]
	IPE	0.64	0.74	0.90	-
Path	(g·m^{-1})	37.5 [a]	29.2 [b]	25.3 [c]	23.7 [c]
	IPE **	1.58	1.23	1.07	-

The same letter in a given row indicates the lack of significant differences between means; * proximity effect index, see Section 2.5; ** border effect index, see Section 2.5.

4. Discussion

So far, the subject of strip intercropping of yellow lupine with other species has not been referenced in scientific literature. Consequently, it is not possible to compare the results presented in this manuscript with findings obtained by other researchers. However, the results regarding the PE of wheat, triticale, barley, and peas on yellow lupine were previously published [28–31]. They were based on an experiment conducted at the same location; similar methodology was also used to determine the effects of strip till on lupine, oats, and triticale [23]. In the discussion, an analogy could be found in regard to the effect on yellow lupine plants grown in MI with other species. Since the cultivation pattern in SI is regular, and MI is characterized by random distribution of different species, the effects of cultivation in MI cannot be compared to those in SI. Hence, there are numerous formulas referencing interactions between different species in regard to the plants' competitive patterns [35], but their application in SI is rather limited.

It has been previously emphasized that yields obtained in MI are more stable in the following years as compared to pure sowing of the species that comprise them. This results from more efficient use of environmental conditions, since one of the most important factors leading to a negative impact on the practice of MI is caused by considerable variability in yield composition in various seasons and different parts of the field. In pure sowing, the yield of yellow lupine is lower than the yield of cereals. This is the effect of biological properties of the species. Consequently, in MI, it is the cereal weight that largely determines the overall size of the yield's total. The dominance of cereals is not only consequence on the mathematical conversion of yield in the pure sowing and participation of components, but it results from mutual interactions that take place in MI during vegetation [36,37]. Generally, cereals are the stronger competitor, hence the lupine yield is only 41%–50% of the yield achieved in pure sowing [5]. A shortage of water intensifies this effect [25,27]. However, it was determined that a single plant of yellow lupine proved to be a stronger competitor than a single plant of oats or triticale. In MI, the asymmetry effect of interspecies competition in favor of cereals results from the fact that cereals are sown at disproportionately higher densities than yellow lupine. Cereals begin to dominate the MI canopy because of their quantitative advantage. Studies have demonstrated the advantage of MI where comparable amounts of oats and lupine were sown, as opposed to MI, where oats comprised the majority of planted species [36–38].

Under the existing experimental conditions, a negative effect of neighboring species in SI on yellow lupine yield was clearly demonstrated. This negative effect translates into SI production results (Figure 4). In the case of two-sided proximity, one must take into account the loss in lupine seed yield ranging from 6.00% (in the neighborhood of peas) to 10.8% (in the neighborhood of barley). However, separating adjacent strips of different species by a non-grown path (in the case of two-side proximity) may lead to an increase in lupine seed yield by 7.8% (Figure 5).

An estimation of the SI yield indicates that the separation of arable strips has an impact on the total yield and changes its structure (Figure 6). It has been proven that the use of a path separating arable strips increases the share of lupine in the total yield. This is most important for SI of yellow lupine with barley. The highest SI yield was obtained for yellow lupine and triticale; but in such a mixture the share of lupine was lower than for SI with other species, which was caused by a relatively high yield of triticale.

Considering that in the total SI yield, the share of lupine exceeded 29%, the above described cultivation method proves to be considerably more advantageous as compared to MI, where yellow lupine's yield share does not usually exceed a dozen percent [5,36,37].

It can be concluded that SI significantly eliminates the problem of the instability of crop composition. It also allows for more efficient habitat exploitation and contributes to the biodiversity of cultivated fields. Appropriate row-width allows for agricultural techniques to be adapted for individual species and facilitates their separate harvesting; compound feed with a desired composition of components can also be obtained. Our results indicate a 10% decrease in lupine yield in SI as a consequence of a two-sided PE in relation to other cereals. In MI, the corresponding decrease in lupine yield reached several dozen percent [5,36,37], indicating a considerable advantage of SI over MI.

The production value of lupine cultivation in combination with other species is evident when the rows of neighboring plants are separated by paths devoid of vegetation. According to our research, the BE contributes increase in the yield of yellow lupine, which, in turn, compensates for the exclusion of path space from production. It should be emphasized that cereals adjacent to the vegetation free path are also being subjected to BE, which results in a several-fold yield increase: up to 40 cm of seeding strip could be affected [26,39]. It should also be noted that the SI production effect depends on rows' geographical orientation: the north-south setting proved to be most favorable [27].

5. Conclusions

The proximity of spring wheat, spring triticale, spring barley, and peas proved to have an adverse effect on the growth, development and yielding of yellow lupine. The unfavorable influence pertained to all the biometric characteristics of yellow lupine. A reduction in value of studied characteristics was noted in the row adjacent to the strip of neighboring species and, for some traits, the negative impact of the PE reached up to the third row of the lupine strip. In the case of SI with 3-m-wide strips, the reduction in lupine yield varied from 6.0% to 10.8%, depending on the neighboring species.

Introduction of a path was demonstrated to have a considerable effect on yellow lupine and resulted in a 7.8% increase in its yield. Furthermore, while taking into account the total yield of SI, the share of yellow lupine was also noticeably higher. Depending on species tested, the share of lupine seeds in the total yield of SI fell in the range of 29%–46%. SI of yellow lupine and triticale proved to be the most beneficial.

Author Contributions: Conceptualization, L.G.; methodology, I.J. and E.W.; investigation—performed the field experiments, L.G. and I.J.; data curation—compiled and analyzed the results, A.W.-P., L.G., I.J. and E.W.; writing—original draft preparation, L.G. and E.W.; review and editing A.W.-P., I.J. and E.W. All authors have read and agreed to the published version of the manuscript.

Funding: This research received no external funding.

Conflicts of Interest: The authors declare no conflict of interest.

References

1. FAOSTAT. Food and Agriculture Organization of the United Nations, Statistics Division. Forestry Production and Trade. Available online: http://www.fao.org/faostat/en/#data/FO (accessed on 4 April 2019).
2. Jul, L.B.; Flengmark, P.; Gylling, M.; Itenov, K. Lupin seed (*Lupinus albus* and *Lupinus luteus*) as a protein source for fermentation use. *Ind. Crop. Prod.* **2003**, *18*, 199–211. [CrossRef]
3. Kaczmarek, S.; Hejdysz, M.; Kubis, M.; Kasprowicz-Potocka, M.; Rutkowski, A. The nutritional value of yellow lupin (*Lupinus luteus* L.) for broilers. *Anim. Feed. Sci. Technol.* **2016**, *222*, 43–53. [CrossRef]
4. Hanczowska, E.; Księżak, J.; Świątkiewicz, M. Efficiency of lupine seed (Lupinus angustifolium and Lupinus luyeus) in sow, piglet and fattener feeding. *Agric. Food Sci.* **2017**, *26*, 1–15. [CrossRef]
5. Księżak, J. Evaluation of mixtures of yellow lupine (*Lupines Luteus* L.) with spring cereals grown for seeds. *Appl. Ecol. Environ. Res.* **2018**, *16*, 1683–1696. [CrossRef]
6. Van der Meer, J.H. The Ecology of Intercropping. *Ecol. Intercropp.* **1989**. [CrossRef]
7. Sánchez, D.G.R.; Silva, J.T.E.; Gil, A.P.; Corona, J.S.S.; Wong, J.A.C.; Mascorro, A.G. Forage yield and quality of intercropped corn and soybean in narrow strips. *Span. J. Agric. Res.* **2010**, *8*, 713. [CrossRef]
8. Głowacka, A. The effect of strip cropping and weed control methods on yields of dent maize, narrow-leafed lupin and oats. *Int. J. Plant Prod.* **2014**, *8*, 505–529. [CrossRef]
9. Gou, F.; van Ittersum, M.; Wang, G.; van der Putten, P.E.; van der Werf, W. Yield and yield components of wheat and maize in wheat–maize intercropping in the Netherlands. *Eur. J. Agron.* **2016**, *76*, 17–27. [CrossRef]
10. Liu, X.; Rahman, T.; Song, C.; Su, B.; Yang, F.; Yong, T.; Wu, Y.; Zhang, C.; Yang, W. Changes in light environment, morphology, growth and yield of soybean in maize-soybean intercropping systems. *Field Crop. Res.* **2017**, *200*, 38–46. [CrossRef]

11. Watson, C.; Atkinson, D.; Gosling, P.; Jackson, L.; Rayns, F. Managing soil fertility in organic farming systems. *Soil Use Manag.* **2006**, *18*, 239–247. [CrossRef]

12. Hassen, A.; Talore, D.G.; Tesfamariam, E.; Friend, M.; Mpanza, T. Potential use of forage-legume intercropping technologies to adapt to climate-change impacts on mixed crop-livestock systems in Africa: A review. *Reg. Environ. Chang.* **2017**, *17*, 1713–1724. [CrossRef]

13. Corre-Hellou, G.; Dibet, A.; Hauggaard-Nielsen, H.; Crozat, Y.; Gooding, M.; Ambus, P.; Dahlmann, C.; von Fragstein, P.; Pristeri, A.; Monti, M.; et al. The competitive ability of pea–barley intercrops against weeds and the interactions with crop productivity and soil N availability. *Field Crop. Res.* **2011**, *122*, 264–272. [CrossRef]

14. Tsubo, M.; Walker, S.; Ogindo, H. A simulation model of cereal–legume intercropping systems for semi-arid regions. *Field Crop. Res.* **2005**, *93*, 23–33. [CrossRef]

15. Sainju, U.M.; Stevens, W.B.; Caesar-Tonthat, T.; Jabro, J.D. Land Use and Management Practices Impact on Plant Biomass Carbon and Soil Carbon Dioxide Emission. *Soil Sci. Soc. Am. J.* **2010**, *74*, 1613–1622. [CrossRef]

16. Doré, T.; Makowski, D.; Malézieux, E.; Munier-Jolain, N.; Tchamitchian, M.; Tittonell, P. Facing up to the paradigm of ecological intensification in agronomy: Revisiting methods, concepts and knowledge. *Eur. J. Agron.* **2011**, *34*, 197–210. [CrossRef]

17. Blum, A. The abiotic stress response and adaptation of triticale—A review. *Cereal Res. Commun.* **2014**, *42*, 359–375. [CrossRef]

18. Wysokiński, A.; Kuziemska, B. The sources of nitrogen for yellow lupine and spring triticale in their intercropping. *Plant Soil Environ.* **2019**, *65*, 145–151. [CrossRef]

19. Xu, Z.; Li, C.; Zhang, C.; Yu, Y.; van der Werf, W.; Zhang, F. Intercropping maize and soybean increases efficiency of land and fertilizer nitrogen use; A meta-analysis. *Field Crop. Res.* **2020**, *246*, 107661. [CrossRef]

20. Lamb, E.G.; Shore, B.H. Water and nitrogen addition differentially impact plant competition in a native rough fescue grassland. *Plant Ecol.* **2006**, *192*, 21–33. [CrossRef]

21. Theunissen, J. Application of Intercropping in Organic Agriculture. *Boil. Agric. Hortic.* **1997**, *15*, 250–259. [CrossRef]

22. Iqbal, N.; Hussain, S.; Ahmed, A.; Yang, F.; Wang, X.C.; Liu, W.G.; Yong, T.W.; Bu, J.B.; Shu, K.; Yang, W.Y.; et al. Comparative analisys of maize-soybean strip intercropping systems: A reniew. *Plant Prod. Sci.* **2019**, *22*, 131–142. [CrossRef]

23. Gałęzewski, L.; Jaskulska, I.; Piekarczyk, M.; Jaskulski, D. Strip intercropping of yellow lupine with oats and spring triticale: Proximity effect. *Acta Sci. Pol. Agric.* **2017**, *16*, 67–75. [CrossRef]

24. Gałęzewski, L. Competition between oat and yellow lupine plants in mixtures of these species. Part I. Intensity of competition depending on soil moisture. *Acta Sci. Pol. Agric.* **2010**, *9*, 37–44.

25. Gałęzewski, L. Competition between oat and yellow lupine plants in mixtures of these species. Part II. Intensity of competition depending on species ratio in mixture. *Acta Sci. Pol. Agric.* **2010**, *9*, 45–52.

26. Gałęzewski, L.; Piekarczyk, M.; Jaskulska, I.; Wasilewski, P. Border effects in the growth of chosen cultivated plant species. *Acta Sci. Pol. Agric.* **2013**, *12*, 3–12.

27. Iragavarapu, T.K.; Randal, G.W. Border effects on yields in a strip-intercropped soyben, corn, and wheat production system. *J. Prod. Agric.* **1996**, *9*, 101–107. [CrossRef]

28. Gałęzewski, L. Proximity effect of spring cereals and legumes in strip intercropping. Part IV. Response of triticale to the proximity of wheat, barley, pea, and yellow lupine. *Acta Sci. Pol. Agric.* **2020**, *19*, 55–67. [CrossRef]

29. Gałęzewski, L.; Jaskulska, I.; Piekarczyk, M. Proximity effect of spring cereals and legumes in strip intercropping. Part I. Response of wheat to the proximity of triticale, barley, pea, and yellow lupine. *Acta Sci. Pol. Agric.* **2018**, *17*, 23–32. [CrossRef]

30. Gałęzewski, L.; Kotwica, K.; Piekarczyk, M. Proximity effect of spring cereals and legumes in strip intercropping. Part II. Response of pea to the proximity of wheat, triticale, barley, and yellow lupine. *Acta Sci. Pol. Agric.* **2018**, *17*, 81–89. [CrossRef]

31. Gałęzewski, L.; Jaskulski, D.; Kotwica, K.; Wasilewski, P. Proximity effect of spring cereals and legumes in strip intercropping. Part III. Response of barley to the proximity of wheat, triticale, pea, and yellow lupine. *Acta Sci. Pol. Agric.* **2018**, *17*, 195–204. [CrossRef]

32. Wikimedia.org. Available online: https://upload.wikimedia.org/wikipedia/commons/b/bd/Poland_in_Europe.svg (accessed on 15 May 2020).

33. Voivodeship. Available online: https://en.wikipedia.org/wiki/Voivodeship (accessed on 15 May 2020).
34. IUSS Working Group WRB. *Word Reference Base for Soil Resources 2014, Update 2015 International Soil Classification System for Naming Soil and Creating Legends for Soil Maps*; World Soil Resources Reports; Springer: Berlin/Heidelberg, Germany, 2015; p. 106.
35. Weigelt, A.; Jolliffe, P. Indices of plant competition. *J. Ecol.* **2003**, *91*, 707–720. [CrossRef]
36. Rudnicki, F.; Gałęzewski, L. Response of oat and yellow lupine on various participation in intercropping and theirs productivity in intercrops. Part I: Reaction of oat and yellow lupine to intercropping. *Zeszyty Problemowe Postępów Nauk Rolniczych* **2007**, *516*, 161–170.
37. Rudnicki, F.; Gałęzewski, L. Response of oat and yellow lupine on various participation in intercropping and theirs productivity in intercrops. Part II. Yielding of oat—Yellow lupine intercrops. *Zeszyty Problemowe Postępów Nauk Rolniczych* **2007**, *516*, 171–179.
38. Rudnicki, F.; Kotwica, K. Comparison of the sffects of spring cereal-legume mixtures with oats, barley or triticale. *Folia Univ. Agric. Stetin. Agric.* **2002**, *228*, 125–130.
39. Niemczyk, H.; Buliński, J. Tramline effect on yield of crops. *Inż. Rol.* **2012**, *2*, 277–286.

 *agriculture*

Article

Phosphorus in Spring Barley and Italian Rye-Grass Biomass as an Effect of Inter-Species Interactions under Water Deficit

Marta K. Kostrzewska *, Magdalena Jastrzębska *, Kinga Treder and Maria Wanic

Department of Agroecosystems, Faculty of Environmental Management and Agriculture, University of Warmia and Mazury in Olsztyn, Plac Łódzki 3, 10-718 Olsztyn, Poland; kinga.treder@uwm.edu.pl (K.T.); maria.wanic@uwm.edu.pl (M.W.)
* Correspondence: marta.kostrzewska@uwm.edu.pl (M.K.K.); jama@uwm.edu.pl (M.J.)

Received: 30 June 2020; Accepted: 3 August 2020; Published: 5 August 2020

Abstract: With global warming, the problem of soil water deficit is growing in Central Europe, including Poland, and the use of catch crops is recommended to mitigate climate changes. This study aimed to determine the influence of water deficit on phosphorus (P) content and accumulation in the above-ground biomass of spring barley and Italian rye-grass growing separately and in the mixture, and on the inter-species interactions between these crops. The study was based on a pot experiment established in accordance with the additive design. The experimental factors were as follows: A. water supply of the plants: an optimal dose and a dose reduced by 50% in relation to the optimal dose, and B. the sowing type: barley sown as a single species, rye-grass sown as a single species, and barley with rye-grass catch crop. Based on the P accumulation in plant biomass, the relative yield of barley and rye-grass, the total relative yield, and the competitive equilibrium index were determined. Water deficit had no effect on the P content in the plants, but it reduced the P accumulation in barley stems, leaves and spikes, as well as in rye-grass stems and leaves, from the emergence to the end of plants' growing period, both when the plants were sown as a single species and as a mixture. Barley was a stronger competitor than rye-grass. Inter-species competition occurred at the stem elongation and heading of barley. The intensification of inter-species competition for P under water deficit conditions should be taken into account when recommending the undersowing of barley with rye-grass for sustainable agriculture.

Keywords: *Hordeum vulgare*; *Lolium multiflorum*; phosphorus; water stress; competition indices; plant development stages

---

1. Introduction

Stress in organisms can be induced by either abiotic or biotic factors [1]. In agro-ecosystems, of all the abiotic factors, drought is the main determinant that limits the development of plants and, consequently, their yielding [2]. Central Europe, including Poland, is located in a temperate climate [3], where the relevance of drought is often underestimated [4]. However, in recent years, with global climate change, the problem of drought has become increasingly serious in the region [4]. Rising atmospheric temperatures have resulted in increasing evapotranspiration rates [4], and seasonal and monthly distributions of precipitation have also been changing [5]. Drying trends were observed for spring and less pronounced for summer, i.e., for a large part of the vegetation period [5]. In general, drought reduces the uptake of minerals and their transport from the roots to the above-ground parts, which affects the rate of plant physiological processes [6]. Tolerance to water stress is a very important feature of plants during drought. Knowledge of species' sensitivity and response to water deficit can be helpful while selecting plants for cultivation, particularly in regions at risk of drought. Numerous studies indicate that resistance to stress is a genotypic trait [7–10].

Catch crops undersown in small grains (main crop) is a form of mixed cropping which offers numerous environmental benefits [11,12], and is considered as a sustainable agricultural practice [12,13]. Catch crop residues left in the field are a significant source of nutrient-rich organic matter [14]. After harvesting the main crop, they serve as ground covers and reduce nitrogen and phosphorus (P) leaching from the soil, as well as direct and indirect greenhouse gas emissions [15–19]. For the latter reason, the cultivation of catch crops is claimed to make an important contribution to climate change mitigation [20,21]. However, underplanted catch crops may compete for nutrients with the main crop [22,23], especially in unfavorable conditions such as water deficit [24]. Italian ryegrass (*Lolium multiflorum*) is one of the most popular catch crops, often undersown in spring barley (*Hordeum vulgare*) [25,26].

Hordeum vulgare is cultivated in many countries worldwide, with the grains intended mainly for animal feed purposes [27]. It is usually cultivated as the main crop. It is ranked fourth, following wheat, rice and maize, in terms of the area under cultivation [28]. Similarly to other cereals, it is also included in mixtures with leguminous plants [29–32], and is used as a protective plant for underplanted catch crops [25,33–35]. The species is distinguished by natural tolerance to drought [36]. This tolerance is determined by early flowering, which ensures optimal pollination, seed development and ripening in an optimal time period. In cereals, the consequences of water deficiency are determined by both the plant's development stage during which the stress occurs [37], and the frequency of drought occurrence during plant development [38].

Lolium multiflorum is a fast-growing annual or perennial grass originating from Europe [39]. *L. multiflorum*, similarly to *L. perenne*, is a valuable fodder plant cultivated in many regions of the world, in both dry and rainy areas [33,40–42]. *L. multiflorum* is cultivated as the main crop, but it can also be cultivated as an underplanted catch crop, similarly to other grasses, papilionaceous plants and their mixtures [15,43,44].

Phosphorus (P) is the second macronutrient after nitrogen, whose deficiency most frequently inhibits plant growth [45]. Poor water availability reduces the uptake of macronutrients from the soil [46,47], and may also affect their content in plants [48–51]. The more severe the drought, the more adverse the effect on the component ratio in the plant [52]. The literature offers many articles on the competition between the cultivated plants, particularly cereals and papilionaceous plants, for habitat resources [53–57]. However, no models have ever been developed to fully explain how plants compete for nutrients under water deficit conditions. Various indicators are used to assess this interaction, e.g., relative yield, relative competitive capacity [58], which are based not only on plant biomass [59], but also on the accumulation of macroelements [23,60]. Relatively more is known about nitrogen accumulation, while there are few studies on phosphorus [61,62]. This study may complement this information.

The study aimed to determine the influence of water deficit on P content and accumulation in the above-ground biomass of spring barley and Italian rye-grass growing separately and in the mixture (rye-grass undersown in barley), and on the inter-species interactions between these crops at different plant development stages.

2. Materials and Methods

2.1. Experimental Design

The study was based on a pot experiment conducted at a greenhouse laboratory of the Faculty of Biology and Biotechnology, University of Warmia and Mazury. The study was carried out on spring barley (Rastik cultivar) and Italian rye-grass (Gaza cultivar).

The experimental factors were as follows:

1. water supply of the plants: optimal (OW), and reduced by 50% in relation to the optimal one (LW),

2. sowing type: barley grown as a single species (BP), rye-grass grown as a single species (RP), barley in a mixture with rye-grass (BM), and rye-grass in a mixture with barley (RM).

An optimal dose of water was determined in a trial experiment, in which plants' irrigation requirements were established based on water loss estimated by daily measurements of pot weight. At the beginning of the trial experiment, the pots with plants were well irrigated, and the soil moisture content was maintained daily by re-watering with the water lost in the previous 24 h. Daily amounts of water supplied to the pots with barley, rye-grass and barley-rye-grass mixture were recorded during the successive growth stages. After finishing the trial experiment, based on the recorded data of water amounts used for daily irrigation, the pattern of plant watering with a higher dose for the proper experiment was established. A higher daily dose of water, common for the three types of sowing, was calculated as an average of barley, rye-grass, and barley-rye-grass mixture requirements at a given stage of plant growth. This dose was dynamic according to the plant development (changeable during vegetation), and it was slightly verified during each growing season. The reduced dose was always equal to one-half of the higher one. At the beginning of each experimental series of the experiment (sowing), the soil moisture was about 20% (measured by time domain reflectometry (TDR) method).

The plants were cultivated on proper brown soil formed from slightly loamy silty sand. The soil had a slightly acidic reaction (pH in 1 M KCl 5.6–6.1), average phosphorus (51–61 mg kg^{-1}), potassium (98–117 mg kg^{-1}) and magnesium (33–42 mg kg^{-1}) content, and an organic carbon content of 7.1–11.1 g kg^{-1}. Each pot, a week before sowing, was filled with 8 kg soil material previously mixed with mineral fertilizers, in a dose of pure component (g pot^{-1}): N–0.5 (urea), P–0.2 (monopotassium phosphate), K–0.45 (potassium sulphate).

Plant kernels were sown into Kick–Brauckmann pots (diameter of 22 cm, depth of 28 cm). When preparing the mixture, the additive pattern was applied, as it assesses the species' interactions at early development stages better than the substitution pattern [63,64]. When single-species sowing was applied, 18 spring barley kernels or 18 Italian rye-grass kernels were sown, while for the mixed-species sowing, 18 spring barley kernels and 18 Italian rye-grass kernels were sown (pure sowing stand). The kernels were distributed using templates at an equal distance from each other over the soil surface, and placed at a depth of 3 cm.

From the kernel sowing to plant harvesting, the temperature at the laboratory was maintained at 20–22 °C. The exception was a 9-day period during the full plant emergence when the temperature was lowered to 6–8 °C to pass the vernalization process.

Three one-year cycles of the experiment were conducted. Each year (cycle), an experiment was set up according to completely randomized design in four replications, and comprised 120 pots: two levels of plant water supply x three levels of sowing type (two species sown separately and in a mixture) x five testing dates x four replications.

2.2. Plant Sampling and Analysis

The phosphorus content in the above-ground biomass of the plants was assayed in five developmental periods, determined by the developmental rhythm of barley sown as a single species and supplied with an optimum water dose. These included (according to BBCH scale): leaf development (10–13), tillering (22–25), stem elongation (33–37), heading (52–55), and ripening (87–91). When barley reached the appropriate stage, the plants were removed from pots (intended for a particular stage), and the shoots were separated from the roots. The material subjected to testing included the above-ground parts of barley and rye-grass plants. The plants were dried in the air and then weighed. For barley, beginning from the stem elongation stage, the shoots were separated into stems and leaves, and from the heading stage, into the spikes as well. For rye-grass, the shoots were separated into the stems and leaves, beginning from the barley stem elongation stage.

The phosphorus content was assayed by the spectrophotometric method (PN-ISO 6491:2000) [65], at the Chemical and Agricultural Research Laboratory in Olsztyn.

2.3. Calculations

The phosphorus accumulation was calculated by multiplying the content of this element in individual parts of the plants by the weight of these organs.

Based on the phosphorus accumulation in the total above-ground biomass of the plants, the relative yield (RY) (Equation (1)) and the total relative yield (RYT) (Equation (2)) were determined [66]:

$$RY_B = Y_{BM}/Y_{BP} \text{ and } RY_R = Y_{RM}/Y_{RP} \tag{1}$$

$$RYT = RY_B + RY_R \tag{2}$$

The relative competitive capacity for phosphorus in mixed seedings was expressed as the competitive equilibrium index (Cb) (Equation (3)) [67]:

$$Cb = \ln[(Y_{BM}/Y_{RM})/(Y_{BP}/Y_{RP})] \tag{3}$$

where: RY_B—relative barley yield, Y_{BM}—barley in mixture yield, Y_{BP}—barley yield for single-species sowing, RY_R—relative rye-grass yield, Y_{RM}—rye-grass in mixture yield, Y_{RP}—rye-grass yield for single-species sowing, ln—natural logarithm.

In the additive pattern, RY < 1 indicates competition, RY > 1 indicates positive interactions, and RY = 1 no interactions. If RYT > 1, this indicates partial complementarity in resource acquisition by the mixture components, positive interactions between species, or incomplete resource acquisition by species in single-species seedings, while if RYT = 2, there are no competitive interactions between species in the mixture, as the resource acquisition by each species in the mixture is the same as in the single-species sowing [60].

The competitive equilibrium index Cb indicates which of the species is more competitive. The experiment calculated Cb for barley in relation to Italian rye-grass. If the species are equal competitors, then Cb = 0, if barley is a better competitor than rye-grass, then Cb > 0, and if rye-grass is a better competitor than barley, then Cb < 0.

2.4. Statistical Analysis

The results were subjected to the analysis of variance (ANOVA). The statistics were calculated separately for each growth stage and for individual plant parts. Homogeneous groups were identified by Duncan's test at $p < 0.05$. Using correlation coefficients, the relationship between the phosphorus accumulation and its content in the organs of plants and their above-ground biomass (the leaves, stems and spikes) was presented. The correlation coefficients were calculated separately for barley and rye-grass, based on the results from all objects (irrespective of the water supply of the plants). It was also checked as to whether the RY and RYT values differ significantly from the unity, and the Cb values from zero, using the t-Student test [60]. Statistical analyses were carried out using the STATISTICA software (data analysis software system), version 12, StatSoft [68].

The results in the tables are means for the three cycles (years) of the experiment.

3. Results

3.1. Phosphorus Content and Accumulation

3.1.1. Spring Barley

The water supply of the plants and the type of sowing had no significant effect on the P content in the above-ground parts of barley throughout the growing period (Table S1).

Water deficit (LW) and the vicinity of rye-grass (BM) reduced P accumulation in the total barley above-ground biomass and individual organs (Table 1). Only at the ripening stage was the P accumulation in barley leaves not affected by the type of sowing. Throughout the cereal vegetation, most P in the above-ground biomass and the organs was accumulated by the plants BP-OW. Water deficit (LW) reduced the P accumulation more than the presence of rye-grass (BM), and the interaction of these factors (BM-LW) resulted in a further reduction in P accumulation.

Table 1. Phosphorus accumulation by barley (mg pot^{-1}).

Growth Stage of Barley	Plant Part	Water Supply		Sowing Type		Water Supply x Sowing Type			
		OW	LW	BP	BM	BP-OW	BM-OW	BP-LW	BM-LW
Leaf development	shoots	5.57 [a]	4.10 [b]	5.22 [a]	4.44 [b]	5.98 [a]	5.15 [b]	4.46 [c]	3.74 [d]
Tillering	shoots	22.47 [a]	10.53 [b]	17.13 [a]	15.87 [b]	23.13 [a]	21.81 [b]	11.13 [c]	9.92 [c]
Stem elongation	shoots	57.01 [a]	22.77 [b]	44.15 [a]	35.64 [b]	62.42 [a]	51.60 [b]	25.88 [c]	19.67 [d]
	leaves	20.87 [a]	10.02 [b]	17.41 [a]	13.48 [b]	23.51 [a]	18.23 [b]	11.32 [c]	8.72 [d]
	stems	36.14 [a]	12.76 [b]	26.74 [a]	22.16 [b]	38.91 [a]	33.37 [b]	14.56 [c]	10.95 [d]
Heading	shoots	68.28 [a]	29.66 [b]	54.37 [a]	43.57 [b]	74.75 [a]	61.82 [b]	33.99 [c]	25.33 [d]
	leaves	18.37 [a]	11.57 [b]	17.47 [a]	12.47 [b]	22.52 [a]	14.22 [b]	12.43 [c]	10.72 [d]
	stems	35.74 [a]	13.99 [b]	26.61 [a]	23.13 [b]	36.57 [a]	34.92 [b]	16.65 [c]	11.34 [d]
	spikes	14.17[a]	4.09 [b]	10.29 [a]	7.98 [b]	15.66 [a]	12.69 [b]	4.91 [c]	3.28 [d]
Ripening	shoots	48.88 [a]	36.58 [b]	46.21 [a]	39.25 [b]	51.85 [a]	45.91 [b]	40.58 [c]	32.59 [d]
	leaves	10.46 [a]	9.42 [b]	10.04 [a]	9.84 [a]	10.40 [ab]	10.51 [a]	9.68 [bc]	9.17 [c]
	stems	20.15 [a]	17.43 [b]	19.86 [a]	17.71 [b]	19.44 [a]	20.85 [a]	20.28 [a]	14.57 [b]
	spikes	18.28 [a]	9.74 [b]	16.31 [a]	11.70 [b]	22.00 [a]	14.55 [b]	10.62 [c]	8.86 [d]

OW—optimal water supply, LW—water supply reduced by 50%; BP—barley as a single species, BM—barley in a mixture with rye-grass; a, b, c, d—homogeneous groups (values followed by the same letters, for each phase and for each part of the plant, within experimental factors and their interactions are not significantly different at $p < 0.05$).

The amount of accumulated P was strongly correlated with the amount of barley above-ground biomass (Table 2). Moreover, a positive correlation was demonstrated between the P accumulation and P content in the vegetative organs of barley during tillering and stem elongation, and between the P accumulation and P content in the barley stems at the heading and ripening stages.

Table 2. Coefficients of correlation (r) between phosphorus accumulation and phosphorus (P) content and the above-ground biomass of plants.

Growth Stage of Barley	Plant Part	Barley		Rye-Grass	
		P Content	Biomass	P Content	Biomass
Leaf development	shoots	0.041	0.977 *	0.291 *	0.960 *
Tillering	shoots	0.618 *	0.944 *	0.584 *	0.993 *
Stem elongation	leaves	0.481 *	0.946 *	0.293 *	0.946 *
	stems	0.602 *	0.968 *	0.229	0.990 *
Heading	leaves	0.116	0.934 *	−0.024	0.963 *
	stems	0.557 *	0.905 *	−0.229	0.980 *
	spikes	0.274	0.942 *	–	–
Ripening	leaves	0.125	0.769 *	0.110	0.850 *
	stems	0.493 *	0.880 *	−0.591 *	0.985 *
	spikes	−0.035	0.994 *	–	–

* r significant at $p < 0.05$.

3.1.2. Italian Rye-Grass

The P content in the vegetative organs of rye-grass throughout the growing period was not affected by the water supply of the plants or the type of sowing (Table S2).

Both water deficit (LW) and the vicinity of barley (RM) reduced the amount of accumulated P in the rye-grass above-ground biomass (Table 3). The strength of the effects of the interaction between water deficiency and barley's competition (RM-LW) on this feature varied at different development stages of the cereal. At the barley leaf development stage, a higher P accumulation was noted in rye-grass RP-OW than RM-OW, in the absence of differences between RP-LW and RM-LW plants. During barley tillering, the reducing effect of water deficit was still noted, but under these conditions, the competitive effect of barley against rye-grass was also observed. From the stem elongation stage until the end of vegetation, the vicinity of barley reduced the P accumulation in rye-grass leaves and stems more than water deficit, and the interaction of stress factors (RM-LW) deepened this reduction.

Table 3. Phosphorus accumulation by rye-grass (mg pot^{-1}).

Growth Stage of Barley	Plant Part	Water Supply		Sowing Type		Water Supply x Sowing Type			
		OW	LW	RP	RM	RP-OW	RM-OW	RP- LW	RM-LW
Leaf development	shoots	0.94 [a]	0.65 [b]	0.85 [a]	0.74 [b]	1.06 [a]	0.82 [b]	0.64 [c]	0.66[c]
Tillering	shoots	9.67 [a]	3.34 [b]	9.54 [a]	3.47 [b]	14.00 [a]	5.33 [b]	5.07 [b]	1.62 [c]
Stem elongation	shoots	28.34 [a]	14.10 [b]	32.71 [a]	9.73 [b]	43.03 [a]	13.64 [c]	22.39 [b]	5.81 [d]
	leaves	14.02 [a]	9.02 [b]	16.50 [a]	6.54 [b]	18.80 [a]	9.23 [c]	14.20 [b]	3.85 [d]
	stems	14.32 [a]	5.08 [b]	16.21 [a]	3.19 [b]	24.23 [a]	4.41 [c]	8.19 [b]	1.96 [d]
Heading	shoots	57.69 [a]	29.82 [b]	65.56 [a]	21.95 [b]	85.22 [a]	30.16 [c]	45.90 [b]	13.74 [d]
	leaves	36.26 [a]	18.98 [b]	40.77 [a]	14.47 [b]	52.56 [a]	19.96 [c]	28.98 [b]	8.99 [d]
	stems	21.43 [a]	10.83 [b]	24.79 [a]	7.47 [b]	32.66 [a]	10.20 [c]	16.92 [b]	4.75 [d]
Ripening	shoots	63.58 [a]	35.71 [b]	70.27 [a]	29.02 [b]	89.6 [a]	37.55 [c]	50.92 [b]	20.49 [d]
	leaves	44.71 [a]	25.71 [b]	50.20 [a]	20.22 [b]	63.64 [a]	25.78 [c]	36.76 [b]	14.66 [d]
	stems	18.82 [a]	10.00 [b]	20.07 [a]	8.75 [b]	25.98 [a]	11.66 [c]	14.16 [b]	5.83 [d]

OW—optimal water supply, LW—water supply reduced by 50%; RP—rye-grass as a single species, RM—rye-grass in a mixture with barley; a, b, c, d—homogeneous groups (values followed by the same letters, for each phase and for each part of the plant, within experimental factors and their interactions are not significantly different at $p < 0.05$).

The P accumulation in the plants was determined by the above-ground biomass of rye-grass (Table 2). Moreover, a positive correlation between the P accumulation and P content in rye-grass vegetative organs until the stem elongation stage and a strong negative correlation between the P accumulation and P content in rye-grass stems during barley ripening were found.

3.2. Competition for Phosphorus

Throughout the growing period, irrespective of the water supply of the plants, there was competition for P between barley and rye-grass ($RY_B < 1$ and $RY_R < 1$) (Table 4). Only during the period of leaf development under water deficit (LW) were no effects of barley on rye-grass noted ($RY_R = 1.04$). A water deficit increased the competition intensity, especially of rye-grass against barley from the stem elongation stage to the end of barley vegetation ($RY_B = 0.75$–0.80) and of barley against rye-grass from the tillering stage to the barley heading stage ($RY_R = 0.26$–0.32). Consequently, during the barley stem elongation and heading stages under water stress conditions, there was full competition between the species (RYT did not differ from 1) (Figure 1). At other development stages, the plants made use of the resource, partially in a complementary manner (RYT > 1).

Table 4. RY values for barley and rye-grass (based on P accumulation) depending on the water supply of the plants.

Growth Stage of Barley	Water Supply			
	OW		LW	
	RY$_B$	RY$_R$	RY$_B$	RY$_R$
Leaf development	0.86 bc*	0.78 a*	0.84 ab*	1.04 a
Tillering	0.94 a	0.38 bc*	0.89 a*	0.32 bc*
Stem elongation	0.83 c*	0.32 c*	0.76 c*	0.26 c*
Heading	0.83 c*	0.35 bc*	0.75 c*	0.30 bc*
Ripening	0.89 b*	0.42 b*	0.80 bc*	0.40 b*

OW—optimal water supply, LW—water supply reduced by 50%; RY$_B$—RY values for barley, RY$_R$—RY values for rye-grass; a, b, c—homogeneous groups (values in the column of the table within species values followed by the same letters are not significantly different at $p < 0.05$); * RY$_B$, RY$_R$, significantly different from 1.0 ($p = 0.05$).

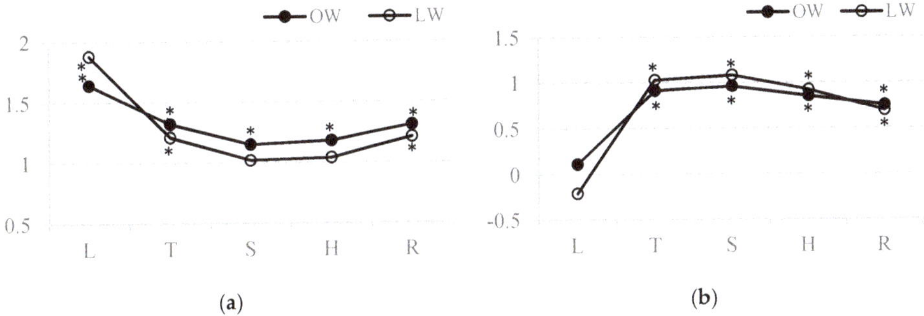

Figure 1. Changes in the RYT (**a**) and Cb (**b**) values for the mixture of barley and rye-grass catch crop during the growth. OW—optimal water supply, LW—water supply reduced 50%; growth stage of barley: L—leaf development, T—tillering, S—stem elongation, H—heading, R—ripening; *—RYT significantly different from 1.0 ($p = 0.05$) and Cb significantly different from 0.0 ($p = 0.05$).

Both under conditions of optimal water supply of the plants and water deficit, the competition of rye-grass against barley was most intense during barley stem elongation and heading. On the other hand, the competition of barley against rye-grass from the tillering to heading stages was the most intense during stem elongation. Barley was a stronger competitor than rye-grass (the Cb index was significantly higher than 0) (Figure 1).

4. Discussion

4.1. Phosphorus Content and Accumulation in Barley

The current study demonstrated no effects of water deficit on the P content in barley above-ground parts throughout the growing period. The results of other studies conducted to date on the effects of water stress on this feature are inconclusive. Both reports of no change [24] and of a reduction in the P content [52,69] of the barley biomass with an increase in water stress were noted.

The type of sowing had no effect on the P content in barley above-ground biomass in the current study. A convergent result was presented by Jastrzębska et al. [24], who examined the effect of red clover as a catch crop accompanying spring barley. On the other hand, Wanic and Michalska [31] demonstrated that mixed barley and pea sowing increased the P content in the cereal above-ground biomass, at the heading and ripening stages.

The reduction in the P accumulation in barley above-ground biomass due to water deficit and the presence of rye-grass was proven in the current study. To compare, Jastrzębska et al. [24] found that mixed spring barley and red clover sowing did not differentiate the amount of P accumulated in the cereal until the heading stage, and at the heading stage, it significantly reduced the amount in the leaves. The current study demonstrated that P accumulation was determined by the biomass produced, and only during the initial period of plant development by P content in the plant. Lower P accumulation under water deficit conditions indicates a smaller biomass. A reduction in barley biomass under drought conditions was confirmed in studies by other authors [52,70–72].

4.2. Phosphorus Content and Accumulation in Rye-Grass

In the current study, the P content in the above-ground stems and leaves of rye-grass was not affected by the water supply of the plants. AbdElgawad et al. [73] also found that drought and high temperatures had no significant effect on the P and other nutrient concentrations in the above-ground biomass of grasses (*Poa pratensis, Lolium perenne*) and legumes (*Lotus corniculatus, Medicago lupulina*). No changes in the P content in rye-grass growing in the vicinity of spring barley were found in the current study either. On the other hand, Høgh-Jensen and Schjoerring [74] demonstrated that P concentration in the dry matter of *Lolium perenne* shoots in the single-species sowing was lower than noted when rye-grass was sown in a mixture with *Trifolium repens*.

There are few studies on the effects of water deficit and competition on P accumulation in the rye-grass biomass. Indirectly, conclusions about it can be drawn based on the biomass, since the P accumulation throughout the growing period had a positive correlation with the above-ground biomass accumulation. Brink et al. [75] also indicates a strong correlation between P uptake and the dry matter of *Lolium multiflorum*, while Burkitt et al. [76] explain such a relationship for *Lolium perenne*. Italian rye-grass is sensitive to water scarcity and primarily responds with poor tillering [77], which can result in lower biomass. The sensitivity of grasses to drought is a feature of the species [78], and even a varietal feature [79], while agronomic and physiological effects of water deficit are determined by the duration of drought [10]. On the other hand, the strength of the effect of a protective (main) crop on the development of catch crop is determined by the protective crop species and the catch crop species. Barley is regarded as a good protective crop for underplanted catch crops [25]. Kuraszkiewicz and Pałys [35] demonstrated that winter rye is a better protective crop than spring barley and oats, and in years with high precipitation, the yields of both fresh and the air-dry above-ground biomass of the catch crop are greater.

4.3. Relative Yields

The nutrient accumulation in plant biomass can result from inter-species interactions such as competition, facilitation and complementarity [74]. For example, a study by Rahetlah et al. [59] demonstrated that Italian rye-grass and spring vetch make complementary use of resources, and that such a mixture could be an alternative to a rye-grass single-crop system, particularly in the dry season. Based on the P accumulation in the plant biomass, expressed as relative yields, it was demonstrated that barley and rye-grass competed for P during the stem elongation and heading stages under water deficit conditions, while at other stages, irrespective of the water supply of the plants, they had a complementary effect (1 < RYT < 2). According to Sobkowicz and Podgórska-Lesiak [60], complementarity always occurs at the early plant development stages in the additive pattern. Before competition or other effects of emerging plants occur, RYT in the additive mixture is always equal to 2, since the yield of each species is the same in the mixture as in the single-species sowing. At subsequent stages of development, barley was more competitive against rye-grass. This is a result of the higher initial barley growth rate following the emergence, which determined its competitive advantage over the slower growing catch crop [30]. A species that grows faster makes use of the necessary growth resources and makes them unavailable to other species [30]. In the current study, the strength of species' competition was changing during vegetation. Both under the conditions of the optimal plants' water supply and of water deficit, barley competed more strongly during the tillering and stem elongation stages and this phenomenon, then started to decrease in intensity.

At the same time, at the barley stem elongation and heading stages, the competition of rye-grass against the cereal (RY_B) began to intensify, which is probably due to the higher growth rate of rye-grass. Similar observations as regards oats as well as rye-grass and vetch catch crop are presented by Paris et al. [80]

5. Conclusions

Water deficit had no effect on the phosphorus content in the above-ground parts of spring barley and Italian rye-grass. This factor reduced phosphorus accumulation in the biomass of barley (the stems, leaves and spikes) and of rye-grass (the stems and leaves) from the emergence to the end of plant growth, both when the plants were cultivated as a single species and in a mixture. Water deficit inhibited the phosphorus accumulation in the barley biomass more than the competition of rye-grass. The competition from barley was, for rye-grass, a stronger factor hindering phosphorus accumulation in the stems and leaves than water deficit. Spring barley was a stronger competitor than rye-grass. Irrespective of the water supply of the plants, the competition intensified until the stem elongation phase. The full competition was noted at the stages of most intense barley development, i.e., during the stem elongation and heading. Underwater deficit inter-species competition for P intensified, which further weakened the P uptake both by barley and rye-grass. This phenomenon should be taken into account when recommending the undersowing of barley with Italian rye-grass for sustainable climate-smart agriculture.

Supplementary Materials: The following are available online at http://www.mdpi.com/2077-0472/10/8/329/s1, Table S1: Phosphorus content of the above-ground biomass of barley (g kg^{-1} dry matter), Table S2: Phosphorus content of the above-ground biomass of rye-grass (g kg^{-1} dry matter).

Author Contributions: Conceptualization, M.W., M.J. and M.K.K.; methodology, M.W., M.J. and M.K.K.; formal analysis, M.K.K. and M.J.; investigation, M.K.K., M.J., M.W. and K.T.; data curation, M.W. and M.K.K.; writing—original draft preparation, M.K.K.; writing—review and editing, M.J.; visualization, M.K.K. and M.J.; funding acquisition, M.K.K., M.J., M.W. and K.T. All authors have read and agreed to the published version of the manuscript.

Funding: This research was funded by the Ministry of Science and Higher Education of Poland, grant number N N310 082836.

Acknowledgments: The authors kindly acknowledge the technical support of Przemysław Makowski from the University of Warmia and Mazury in Olsztyn.

Conflicts of Interest: The authors declare no conflict of interest.

References

1. Suzuki, N.; Rivero, R.M.; Shulaev, V.; Blumwald, E.; Mittler, R. Abiotic and biotic stress combinations. *New Phytol.* **2014**, *203*, 32–43. [CrossRef] [PubMed]
2. Valliyodan, B.; Nguyen, H.T. Understanding regulatory networks and engineering for enhanced drought tolerance in plants. *Curr. Opin. Plant Biol.* **2006**, *9*, 189–195. [CrossRef] [PubMed]
3. Michalak, D. Adapting to climate change and effective water management in Polish agriculture—At the level of government institutions and farms. *Ecohydrol. Hydrobiol.* **2020**, *20*, 134–141. [CrossRef]
4. Hänsel, S.; Ustrnul, Z.; Łupikasza, E.; Skalak, P. Assessing seasonal drought variations and trends over Central Europe. *Adv. Water Resour.* **2019**, *127*, 53–75. [CrossRef]
5. Szwed, M. Variability of precipitation in Poland under climate change. *Theor. Appl. Climatol.* **2019**, *135*, 1003–1015. [CrossRef]
6. da Silva, E.C.; Nogueira, R.; da Silva, M.A.; de Albuquerque, M.B. Drought stress and plant nutrition. *Plant Stress* **2011**, *5*, 32–41.
7. Crush, J.R.; Easton, H.S.; Waller, J.E.; Hume, D.E.; Faville, M.J. Genotypic variation in patterns of root distribution, nitrate interception and response to moisture stress of a perennial ryegrass (Lolium perenne L.) mapping population. *Grass Forage Sci.* **2007**, *62*, 265–273. [CrossRef]
8. Samarah, N.H.; Alqudah, A.M.; Amayreh, J.A.; McAndrews, G.M. The effect of late-terminal drought stress on yield components of four barley cultivars. *J. Agron. Crop Sci.* **2009**, *195*, 427–441. [CrossRef]

9. Pecio, A.; Wach, D. Grain yield and yield components of spring barley genotypes as the indicators of their tolerance to temporal drought stress. *Pol. J. Agron.* **2015**, *21*, 19–27.

10. Cyriac, D.; Hofmann, R.W.; Stewart, A.; Sathish, P.; Winefield, C.S.; Moot, D.J. Intraspecific differences in long-term drought tolerance in perennial ryegrass. *PLoS ONE* **2018**, *13*, e194977. [CrossRef]

11. Aronsson, H.; Hansen, E.M.; Thomsen, I.K.; Liu, J.; Øgaard, A.F.; Känkänen, H.; Ulén, B. The ability of cover crops to reduce nitrogen and phosphorus losses from arable land in southern Scandinavia and Finland. *J. Soil Water Conserv.* **2016**, *71*, 41–55. [CrossRef]

12. Żuk-Gołaszewska, K.; Wanic, M.; Orzech, K. The role of catch crops in field plant production—A review. *J. Elem.* **2019**, *24*, 575–587. [CrossRef]

13. Wanic, M.; Zuk-Golaszewska, K.; Orzech, K. Catch crops and the soil environment—A review of the literature. *J. Elem.* **2019**, *24*, 31–45. [CrossRef]

14. Arlauskiene, A.; Maiksteniene, S.; Slepetiene, A. Application of environmental protection measures for clay loam cambisol used for agricultural purposes. *J. Environ. Eng. Landsc. Manag.* **2011**, *19*, 71–80. [CrossRef]

15. Doltra, J.; Olesen, J.E. The role of catch crops in the ecological intensification of spring cereals in organic farming under Nordic climate. *Eur. J. Agron.* **2013**, *44*, 98–108. [CrossRef]

16. Hansen, E.M.; Eriksen, J.; Vinther, F.P. Catch crop strategy and nitrate leaching following grazed grass-clover. *Soil Use Manag.* **2007**, *23*, 348–358. [CrossRef]

17. Känkänen, H.; Eriksson, C. Effects of undersown crops on soil mineral N and grain yield of spring barley. *Eur. J. Agron.* **2007**, *27*, 25–34. [CrossRef]

18. Malcolm, B.J.; Moir, J.L.; Cameron, K.C.; Di, H.J.; Edwards, G.R. Influence of plant growth and root architecture of Italian ryegrass (Lolium multiflorum) and tall fescue (Festuca arundinacea) on N recovery during winter. *Grass Forage Sci.* **2015**, *70*, 600–610. [CrossRef]

19. Abdalla, M.; Hastings, A.; Cheng, K.; Yue, Q.; Chadwick, D.; Espenberg, M.; Truu, J.; Rees, R.M.; Smith, P. A critical review of the impacts of cover crops on nitrogen leaching, net greenhouse gas balance and crop productivity. *Glob. Chang. Biol.* **2019**, *25*, 2530–2543. [CrossRef]

20. Guardia, G.; Aguilera, E.; Vallejo, A.; Sanz-Cobena, A.; Alonso-Ayuso, M.; Quemada, M. Effective climate change mitigation through cover cropping and integrated fertilization: A global warming potential assessment from a 10-year field experiment. *J. Clean. Prod.* **2019**, *241*. [CrossRef]

21. Bert, S.; Bas, J.; Wiepie, H.; Wil, H.; Luis, A.J.; Jonas, K. *Adoption of Cover Crops for Climate Change Mitigation in the EU*; Joint Research Centre (Seville Site): Luxembourg, 2019.

22. Sobkowicz, P. Uptake and use of nitrogen in the cultivation of spring barley with undersown Persian clover and serradella. *Probl. Inż. Rol.* **2009**, *2*, 99–106.

23. Wanic, M.; Myśliwiec, M.; Orzech, K.; Michalska, M.; Denert, M. Competition for nitrogen, phosphorus, potassium and magnesium between spring wheat and persian clover depending on the density of plants. *J. Elem.* **2017**, *22*, 1081–1093. [CrossRef]

24. Jastrzebska, M.; Kostrzewska, M.K.; Wanic, M.; Makowski, P.; Treder, K. Phosphorus content in spring Barley and red clover plants in pure and mixed sowing. *Acta Sci. Pol. Agric.* **2015**, *14*, 21–32.

25. Kunelius, H.T.; McRae, K.B.; Dürr, G.H.; Fillmore, S.A.E. Management of Italian and perennial ryegrasses for seed and forage production in crop rotations. *J. Agron. Crop Sci.* **2004**, *190*, 130–137. [CrossRef]

26. Salonen, J.; Ketoja, E. Undersown cover crops have limited weed suppression potential when reducing tillage intensity in organically grown cereals. *Org. Agric.* **2020**, *10*, 107–121. [CrossRef]

27. Giraldo, P.; Benavente, E.; Manzano-Agugliaro, F.; Gimenez, E. Worldwide research trends on wheat and barley: A bibliometric comparative analysis. *Agronomy* **2019**, *9*, 352. [CrossRef]

28. FAO (Food and Agriculture Organization of the United Nations). Crops Barley; 2018. Available online: http://www.fao.org/faostat/en/#data/QC (accessed on 7 February 2020).

29. Agegnehu, G.; Ghizaw, A.; Sinebo, W. Yield performance and land-use efficiency of barley and faba bean mixed cropping in Ethiopian highlands. *Eur. J. Agron.* **2006**, *25*, 202–207. [CrossRef]

30. Sobkowicz, P. Shoot and root competition between spring triticale and field beans during early growth. *Acta Sci. Pol. Agric.* **2005**, *4*, 117–126.

31. Wanic, M.; Michalska, M. The influence of competition between spring barley and field pea on content of macroelements in different parts of the plants. *Fragm. Agron.* **2009**, *26*, 162–174.

32. Pappa, V.A.; Rees, R.M.; Walker, R.L.; Baddeley, J.A.; Watson, C.A. Legumes intercropped with spring barley contribute to increased biomass production and carry-over effects. *J. Agric. Sci.* **2012**, *150*, 584–594. [CrossRef]

33. Andersen, A.; Olsen, C.C. Rye grass as a catch crop in spring barley. *Acta Agric. Scand. Sect. B Soil Plant Sci.* **1993**, *43*, 218–230. [CrossRef]

34. Ohlander, L.; Bergkvist, G.; Stendahl, F.; Kvist, M. Yield of catch crops and spring barley as affected by time of undersowing. *Acta Agric. Scand. Sect. B Soil Plant Sci.* **1996**, *46*, 161–168. [CrossRef]

35. Kuraszkiewicz, R.; Palys, E. The influence of cover crops on the yield of aboveground parts of undersown crops. *Ann. Univ. Mariae Curie Sklodowska Sectio E Agric. (Poland)* **2005**, *57*, 105–112.

36. Gürel, F.; Öztürk, Z.N.; Uçarlı, C.; Rosellini, D. Barley genes as tools to confer abiotic stress tolerance in crops. *Front. Plant Sci.* **2016**, *7*. [CrossRef] [PubMed]

37. Akram, M. Growth and yield components of wheat under water stress of different growth stages. *Bangladesh J. Agric. Res.* **2011**, *36*, 455–468. [CrossRef]

38. Ding, J.; Huang, Z.; Zhu, M.; Li, C.; Zhu, X.; Guo, W. Does cyclic water stress damage wheat yield more than a single stress? *PLoS ONE* **2018**, *13*, e0195535. [CrossRef]

39. *The Biology of Lolium multiflorum Lam (Italian ryegrass), Lolium perenne (perenial ryegrass) and Lolium arundinaceum (Schreb) Darbysh (tall fescue)*; Department of Primary Industries: Melbourne, VIC, Australia, 2008; p. 83.

40. Varella, A.C.; Carassai, I.J.; Baldissera, T.C.; Nabinger, C.; Lustosa, S.B.C.; Moraes, A.; Teixeira, S.J.; Vargas, A.S.; Radin, B. Annual ryegrass dry matter yield and nitrogen responses to fertiliser N applications in southern Brazil. *Agron. N. Z.* **2010**, *40*, 33–42.

41. Mandić, V.; Simić, A.; Vučković, S.; Stanisavljević, R.; Tomić, Z.; Bijelić, Z.; Krnjaja, V. Management practices effect on seed features of Italian ryegrass following storage period. *Biotechnol. Anim. Husb.* **2014**, *30*, 145–152. [CrossRef]

42. Kim, M.; Peng, J.L.; Sung, K. Causality between climatic and soil factors on Italian ryegrass yield in paddy field via climate and soil big data. *J. Anim. Sci. Technol.* **2019**, *61*, 324–332. [CrossRef]

43. Olesen, J.E.; Hansen, E.M.; Askegaard, M.; Rasmussen, I.A. The value of catch crops and organic manures for spring barley in organic arable farming. *Field Crop. Res.* **2007**, *100*, 168–178. [CrossRef]

44. Płaza, A.; Ceglarek, F.; Gąsiorowska, B.; Królikowska, M.A. The yielding and chemical composition of undersown crops. *Fragm. Agron.* **2009**, *26*, 93–99.

45. Schachtman, D.P.; Reid, R.J.; Ayling, S.M. Phosphorus Uptake by Plants: From Soil to Cell. *Plant Physiol.* **1998**, *116*, 447–453. [CrossRef] [PubMed]

46. Nawaz, F.; Ahmad, R.; Waraich, E.A.; Naeem, M.S.; Shabbir, R.N. Nutrient uptake, physiological responses, and yield attributes of wheat (triticum aestivum l.) exposed to early and late drought stress. *J. Plant Nutr.* **2012**, *35*, 961–974. [CrossRef]

47. Suriyagoda, L.D.B.; Ryan, M.H.; Renton, M.; Lambers, H. Plant responses to limited moisture and phosphorus availability: A meta-analysis. In *Advances in Agronomy*; Academic Press Inc.: San Diego, CA, USA, 2014; Volume 124, pp. 143–200.

48. Baligar, V.C.; Fageria, N.K.; He, Z.L. Nutrient use efficiency in plants. *Commun. Soil Sci. Plant Anal.* **2001**, *32*, 921–950. [CrossRef]

49. Farahani, S.M.; Chaichi, M.R.; Mazaheri, D.; Afshari, R.T.; Savaghebi, G. Barley grain mineral analysis as affected by different fertilizing systems and by drought stress. *J. Agric. Sci. Technol.* **2011**, *13*, 315–326.

50. Gonzalez-Dugo, V.; Durand, J.L.; Gastal, F.; Bariac, T.; Poincheval, J. Restricted root-to-shoot translocation and decreased sink size are responsible for limited nitrogen uptake in three grass species under water deficit. *Environ. Exp. Bot.* **2012**, *75*, 258–267. [CrossRef]

51. Brodowska, M.S.; Filipek, T.; Kurzyna-Szklarek, M. Content of magnesium and calcium in cultivated plants depending on various soil supply with nitrogen, potassium, magnesium and sulfur. *J. Elem.* **2017**, *22*, 1167–1177. [CrossRef]

52. Bista, D.R.; Heckathorn, S.A.; Jayawardena, D.M.; Mishra, S.; Boldt, J.K. Effects of drought on nutrient uptake and the levels of nutrient-uptake proteins in roots of drought-sensitive and -tolerant grasses. *Plants* **2018**, *7*, 28. [CrossRef]

53. Dhima, K.V.; Lithourgidis, A.S.; Vasilakoglou, I.B.; Dordas, C.A. Competition indices of common vetch and cereal intercrops in two seeding ratio. *Field Crop. Res.* **2007**, *100*, 249–256. [CrossRef]

54. Mariotti, M.; Masoni, A.; Ercoli, L.; Arduini, I. Above- and below-ground competition between barley, wheat, lupin and vetch in a cereal and legume intercropping system. *Grass Forage Sci.* **2009**, *64*, 401–412. [CrossRef]

55. Eskandari, H.; Ghanbari, A. Effect of different planting pattern of wheat (Triticum aestivum) and bean (Vicia faba) on grain yield, dry matter production and weed biomass. *Not. Sci. Biol.* **2010**, *2*, 111–115. [CrossRef]

56. Alaru, M.; Talgre, L.; Luik, A.; Tein, B.; Eremeev, V.; Loit, E. Barley undersown with red clover in organic and conventional systems: Nitrogen aftereffect on legume growth. *Zemdirbyste* **2017**, *104*, 131–138. [CrossRef]

57. Darch, T.; Giles, C.D.; Blackwell, M.S.A.; George, T.S.; Brown, L.K.; Menezes-Blackburn, D.; Shand, C.A.; Stutter, M.I.; Lumsdon, D.G.; Mezeli, M.M.; et al. Inter- and intra-species intercropping of barley cultivars and legume species, as affected by soil phosphorus availability. *Plant Soil* **2018**, *427*, 125–138. [CrossRef] [PubMed]

58. Weigelt, A.; Jolliffe, P. Indices of plant competition. *J. Ecol.* **2003**, *91*, 707–720. [CrossRef]

59. Rahetlah, V.B.; Randrianaivoarivony, J.M.; Andrianarisoa, B.; Razafimpamoa, L.H.; Ramalanjaona, V.L. Yields and quality of Italian ryegrass (Lolium multiflorum) and Common Vetch (Vicia sativa) grown in monocultures and mixed cultures under irrigated conditions in the Highlands of Madagascar. *Sustain. Agric. Res.* **2013**, *2*. [CrossRef]

60. Sobkowicz, P.; Podgórska-Lesiak, M. Assessment of barley interactions in mixture with triticale or field pea affected by nitrogen fertilizer rate. *Fragm. Agron.* **2009**, *26*, 115–126.

61. Hinsinger, P.; Betencourt, E.; Bernard, L.; Brauman, A.; Plassard, C.; Shen, J.; Tang, X.; Zhang, F. P for two, sharing a scarce resource: Soil phosphorus acquisition in the rhizosphere of intercropped species. *Plant Physiol.* **2011**, *156*, 1078–1086. [CrossRef]

62. Li, C.; Dong, Y.; Li, H.; Shen, J.; Zhang, F. Shift from complementarity to facilitation on P uptake by intercropped wheat neighboring with faba bean when available soil P is depleted. *Sci. Rep.* **2016**, *6*. [CrossRef]

63. Snaydon, R.W. Replacement or additive designs for competition studies? *J. Appl. Ecol.* **1991**, *28*, 930–946. [CrossRef]

64. Semere, T.; Froud-Williams, R.J. The effect of pea cultivar and water stress on root and shoot competition between vegetative plants of maize and pea. *J. Appl. Ecol.* **2001**, *38*, 137–145. [CrossRef]

65. *Pasze—Oznaczanie Zawartości Fosforu—Metoda Spektrometryczna*; Polski Komitet Normalizacyjny: Warszawa, Polska, 2000; p. 9.

66. De Wit, C.T.; Van den Bergh, J.P. Competition between herbage plants. *J. Agric. Sci.* **1965**, *13*, 212–221.

67. Wilson, J.B. Shoot competition and root competition. *J. Appl. Ecol.* **1988**, *25*, 279–296. [CrossRef]

68. StatSoft, I. *Statistica (Data Analysis Software System), Version 12*; Statsoft Inc.: Tulsa, OK, USA, 2014.

69. Kristoffersen, A.Ø.; Riley, H. Effects of soil compaction and moisture regime on the root and shoot growth and phosphorus uptake of barley plants growing on soils with varying phosphorus status. *Nutr. Cycl. Agroecosyst.* **2005**, *72*, 135–146. [CrossRef]

70. Alghabari, F.; Ihsan, M.Z. Effects of drought stress on growth, grain filling duration, yield and quality attributes of barley (Hordeum vulgare L.). *Bangladesh J. Bot.* **2018**, *47*, 421–428. [CrossRef]

71. Ali, A.H.; Mansorr, H.N. Effect of imposed water stress at certain growth stages on growth and yield of barley grown under different planting patterns. *Plant Arch.* **2018**, *18*, 1735–1744.

72. Boudiar, R.; Casas, A.M.; Gioia, T.; Fiorani, F.; Nagel, K.A.; Igartua, E. Effects of low water availability on root placement and shoot development in landraces and modern barley cultivars. *Agronomy* **2020**, *10*, 134. [CrossRef]

73. AbdElgawad, H.; Peshev, D.; Zinta, G.; Van Den Ende, W.; Janssens, I.A.; Asard, H. Climate extreme effects on the chemical composition of temperate grassland species under ambient and elevated CO2: A comparison of fructan and non-fructan accumulators. *PLoS ONE* **2014**, *9*, e92044. [CrossRef]

74. Høgh-Jensen, H.; Schjoerring, J.K. Interactions between nitrogen, phosphorus and potassium determine growth and N2-fixation in white clover and ryegrass leys. *Nutr. Cycl. Agroecosyst.* **2010**, *87*, 327–338. [CrossRef]

75. Brink, G.E.; Pederson, G.A.; Sistani, K.R.; Fairbrother, T.E. Forages: Uptake of selected nutrients by temperate grasses and legumes. *Agron. J.* **2001**, *93*, 887–890. [CrossRef]

76. Burkitt, L.L.; Turner, L.R.; Donaghy, D.J.; Fulkerson, W.J.; Smethurst, P.J.; Roche, J.R. Characterisation of phosphorus uptake by perennial ryegrass (Lolium perenne L.) during regrowth. *N. Z. J. Agric. Res.* **2009**, *52*, 195–202. [CrossRef]

77. Norris, I.B.; Thomas, H. Recovery of ryegrass species from drought. *J. Agric. Sci.* **1982**, *98*, 623–628. [CrossRef]
78. Jerónimo, P.A.; Hrabě, F.; Knot, P.; Kvasnovský, M. Evaluation of suitability of grass species for dry conditions (water stress). *Acta Univ. Agric. Silvic. Mendel. Brun.* **2014**, *62*, 953–960. [CrossRef]
79. Staniak, M. The impact of drought stress on the yields and food value of selected forage grasses. *Acta Agrobot.* **2016**, *69*. [CrossRef]
80. Paris, W.; Marchesan, R.; Cecato, U.; Martin, T.N.; Ziech, M.F.; Borges, G.D.S. Dynamics of yield and nutritional value for winter forage intercropping. *Acta Sci. Anim. Sci.* **2012**, *34*, 109–115. [CrossRef]

Article

Evaluation of the Productivity of New Spring Cereal Mixture to Optimize Cultivation under Different Soil Conditions

Danuta Leszczyńska [1],*, Agnieszka Klimek-Kopyra [2],* and Krzysztof Patkowski [3]

[1] Department of Cereal Crop Production, Institute of Soil Science and Plant Cultivation—State Research Institute, Czartoryskich 8, 24-100 Puławy, Poland
[2] Department of Agroecology and Plant Production, University of Agriculture in Cracow, Mickiewicza 21, 31-120 Kraków, Poland
[3] Institute of Animal Breeding and Biodiversity Conservation, University of Life Sciences in Lublin, Akademicka 13, 20-950 Lublin, Poland; krzysztof.patkowski@up.lublin.pl
* Correspondence: leszcz@iung.pulawy.pl (D.L.); agnieszka.klimek@urk.edu.pl (A.K.-K.)

Received: 30 June 2020; Accepted: 7 August 2020; Published: 9 August 2020

Abstract: The aim of the study was to evaluate grain yields, protein yields, and net metabolic energy yields of different combinations of spring types of barley, oat, and wheat arranged in 10 mixtures and grown under different soil types. Naked cultivars of barley and oat were used. The three-year field experiment was conducted at the Agricultural Advisory Centre in Szepietowo, Poland. The study showed that the major factor determining yields of the mixtures was soil quality. Within the better soil (Albic Luvisols), the highest yield was achieved by a mixture of covered barley and wheat and by a mixture of covered barley with covered oats and wheat, but only in treatments with lower sowing density. Moreover, on the better soil, significantly higher protein yields were obtained for mixtures of barley (covered or naked grains) with wheat as compared to the mixture of covered barley with covered oats, or the mixture of covered barley with naked oats and wheat. The highest yields of net metabolic energy, regardless of soil type, were obtained from a mixture of naked barley with wheat, while the lowest from a mixture of covered barley with naked oats and wheat. Mixed sowings increase biodiversity of canopies, which allows a better use of production space. They also increase health and the productivity of plants.

Keywords: spring cereal mixtures; grain yield; protein yield; metabolic energy yield; differentiations of cereal mixture; sustainable agriculture

---

1. Introduction

A growing demand for food in developed and developing countries as well as natural disasters such as drought, disease, and pests are becoming a major challenge for agricultural production in the 21st century. Today, purely species-specific crops dominate the cultivation of cereals. The opposite option to pure sowing of cereal species may be cereal mixtures, mainly interspecies, which are currently estimated to account for 1% of this group of agricultural crops [1]. Cereals and cereal-and-legume mixtures are an essential link in the transition of sustainable agriculture and organic farming [2]. In Central Europe, a disturbing trend is the high percentage of cereals in the sowing structure, which results in a succession of cereal crops for several years. Moreover, each simplification in the tillage system increases the disturbance of biological balance in the agricultural environment. The dominance of one cereal species, or of one cultivar within a species in a given area, promotes the development of pathogens, which causes a decrease in yields. Cereal mixtures maintain better plant health by increasing the biodiversity of the canopy [3,4].

The increase of plant productivity based on biodiversity is conditioned by more effective use of the interrelationships among plants in a mixed crop [5,6]. One of the concepts for increasing plant productivity in this cultivation system is to optimize plant species selection in mixed sowings to make complementary use of available space, water, and nutrients [7,8]. This concept was presented by Li et al. [9] who claimed that the complementary effect is associated with a better use of space by one of the components of the mixture, which has not been fully utilized by other components less tolerant to the given habitat conditions. It can be done by using plant architecture as a strategy to allow one member of a mix to capture sunlight that would otherwise be unused. According to Li et al. [9], this concept is considered a spatial complementarity, and the phenomenon of competitiveness in that concept is defined as the properties of the species that are characterized by faster development and better control of the space, which limits the development of other components in mixed sowing [10–12]. In mixed sowing, lower weed infestation, poorer pest infestation, and better resistance to lodging are observed due to the production of lower and more flexible stems [13,14]. Different species in mixtures better penetrate the soil thanks to their different root systems and enable a more efficient use of fertilization, which can be applied in smaller doses [3]. A two-species mixed sowing, consisting of species of different crop groups, is not a common practice in mechanised systems due to higher labour input, mainly during sowing and harvesting, as well as due to the instability in yields due to weather conditions [3,13,15]. An alternative agrotechnical solution is to compose mixtures or mixes within one group of plants, e.g., cereals. Cereal species mixtures can increase the intra-species diversity of the cropping system diversity by increasing genetic diversity in the canopy. Such use of intra-species diversity is well suited to mechanised systems that are designed to manage a single species, as it can provide benefits from reduced disease, weed, and insect pressure as well as improve yield level and quality per hectare [16,17]. In large farms, where the share of cereals in the farming systems often reaches 75% and where genetic uniformity lead to a biological imbalance in the fields, the use of multispecies cereal mixtures becomes a desirable solution [18].

In central Europe, there is a large variation in soil quality, which is a major problem in terms of increasing the productivity of cereal crops. To minimize the impact of soil variability on yield, mixed sowing is promoted. Hong et al. [6] have shown that, on small and large farms, the yield is determined by the sowing method. The larger the field area, the greater the overall variability of soil quality, which favours yields of mixtures compared to sole crop/pure stand, due to the dominance in the canopy of this cereal species for which the soil characteristics are appropriate. Cultivating naked cultivars of spring barley and oats in pure sowings, due to lower yields, is less economical than pure sowing of the covered forms of these cereals. It is recommended to cultivate naked cultivars of these cereals in mixtures intended as fodder on one's own farm due to better quality of grains of naked forms [19–21]. Mixed sowing increases biodiversity in the fields and contributes to the sustainability of crop production [3,7]. On large farms, the percentage of cereals in the monoculture is high, and crop simplifications often lead to an imbalance in biological diversity. Growing cereals in mixtures contributes to improving crop health. In addition, mixtures are more tolerant of unfavourable weather conditions and varied habitat conditions across the field than the tolerance of pure crop sowings of cereals. Due to lower susceptibility to climatic factors of limiting nature (shortage of precipitation, large temperature fluctuations), barley exhibits higher yield reliability than other spring cereals. The advantage of barley (as a component of the mixture) is the greatest resistance to drought among spring cereals due to a lower transpiration coefficient and high root suction power. Barley as a component of the mixture, in comparison with other cereal species, is very sensitive to soil acidification [22].

The inclusion of naked barley and naked oats into a mixture increases the protein and fat contents in the grains of the mixture, which contributes to their better forage value. It is advantageous to cultivate naked cereal cultivars in mixtures that are intended for fodder because of better quality of naked grain forms [23]. In the absence of the husk, the metabolised energy of the naked oat kernel can be comparable to or higher than that of wheat [24]. Naked oat kernels have also been shown to have a higher content of metabolised energy, lipids, linoleic acid, protein, essential amino acids, and starch than husked oat cultivars [25,26]. These characteristics make naked oats potentially more suitable as a feed source than other cereals particularly for poultry (MacLeod et al., 2008). Oat grain has a number of nutritional benefits compared to other cereals [27]. It has a high lipid content compared to wheat and barley, which comprises principally unsaturated oleic and linoleic fatty acids as well as high concentrations of the amino acids' lysine, methionine, and cysteine [23]. For human nutrition, they are a source of soluble fibre and β-glucans, which both can have positive effects on health [23,28].

The oats are characterized by high phytosanitary properties and, thus, they can reduce the infestation of the mixture canopy by fungal diseases. A novelty of the study lies in the comparison of different spring species' composition of mixtures when taking into account new type cultivars (naked vs. covered) of barley and oats. This creates new possibilities to increase the yield and quality of grains without increasing the expenditure on chemical means of grain production under conditions of sustainable agriculture. In the light of unpredictable environmental variation factors, the great impediment is choosing the right cultivar or cultivar mixture. Ločmele et al. [29] highlighted that it is unclear how many cultivars and which type of cultivar should be used to compose the mixture.

The aim of the study was to compare the yield, protein yield, and net metabolic energy of different variants of spring cereal mixtures with the share of naked cultivars of spring barley and oats at different sowing densities, depending on soil quality. The scientific hypothesis assumed that grain yield differentiation within cereal mixture variants would be different than protein and metabolic energy yield diversification due to the lower grain yield, but a higher content of protein and metabolic energy in the grains of naked forms of barley and oats as well as wheat. A higher grain yield is expected from a mixture of hulled forms of spring barley and oats. However, the yield of protein and metabolic energy of mixtures with the share of naked forms of barley, oats, and wheat, should be similar to that of covered forms.

2. Materials and Methods

The field experiment with different combinations of spring cereal mixtures was conducted as part of field experimentation of Podlaskie Agricultural Advisory Centre in Szepietowo (AAC), Poland (52°52′, 22°32′), in the years 2013–2015. Two, two-factorial field experiments (with the same treatments) on different types of soils were performed. The experiments were conducted on better-quality soil: Albic Luvisols (developed in loamy sand on loam), and on poorer quality soil: Haplic Arenosols (developed in loamy sand on sand), (Table 1). The first (random) factor was study years while the second factor was 10 sowing combinations. Mixture variants differed in the species composition—hulled covered barley (*Hordeum vulgare* L. cv. Skarb), naked barley (cv. Gawrosz), covered oats (*Avena sativa* L. cv. Krezus), naked oats (*Avena nuda* L. cv. Nagus), and wheat (*Triticum aestivum* L. cv. Nawra), as well as in sowing density (Table 2). The field experiment was carried out in four replications and the size of a single plot was 15 m² (length 10 m, width 1.5 m). Each plot consisted of 12 rows with a row-spacing of 12.5 cm. Grains were treated with thiram (37.5%) and carboxin (37.5%) (Oxafun T) and sowed using an Oyjord plot drill to a depth of 4 cm. After that, the sowing plots were harrowed, using a light harrow. The cultivation of barley, wheat, and oats in pure sowing was well recognized in earlier studies by the authors [22,30]. Leszczyńska and Noworolnik [22] proved that productivity of covered oat and barley is related to soil quality. In rich soil (clay soil), oat yields at 4.9 t ha^{-1} and barley on 5.36 t ha^{-1}, while in pure (sandy soil) oat yields at 4.46 t ha^{-1} and barley on 3.87 t ha^{-1}. Szmigiel and Oleksy [30] indicated that cultivation of covered cultivars of oat or barley in pure sowing was more beneficial than cultivation of naked cultivars of this species. The covered oat cv. 'Chwat' yielded 58% higher (5.85 t ha^{-1}) compared to naked cv. 'Akt' (3.71 t ha^{-1}) while covered barley cv. 'Rodos' yielded 11% higher (4.41 t ha^{-1}) compared to naked cv. 'Rastik' (3.97 t ha^{-1}). Zając et al. [18] indicated that wheat in pure sowing was yielded at 8 t ha^{-1}.

Table 1. Nutrient content and soil pH of the experimental field in AAC Szepietowo in 2013–2015.

Specification	Soil Characteristics According to New Soil Classification					
Soil Type	Albic Luvisols			Haplic Arenosol		
Year	2013	2014	2015	2013	2014	2015
pH in KCl	5.9	5.9	6.8	4.9	5.1	4.9
P_2O_5 mg/100 g soil	15.4	8.7	20.7	11.0	9.0	13.4
K_2O mg/100 g soil	13.5	11.7	17.0	13.9	14.4	10.5
Mg mg/100 g soil	7.8	7.8	5.7	3.4	4.8	4.0

Table 2. Experimental scheme with spring cereal mixtures.

Treatment	Mixture Composition	* Sowing Rate of Cereals Seed Number Per 1 m^{-2}
I	Covered barley + covered oats (CB + CO)	160 + 300
II	Covered barley + wheat (CB + W)	160 + 290
III	Covered barley + covered oats + wheat (CB + CO + W)	107 + 200 + 193
IV	Naked barley + wheat (NB + W)	160 + 290
V	Covered barley + naked oats + wheat (CB + NO + W)	107 + 220 + 193
VI	Covered barley + covered oats (CB + CO)	136 + 255
VII	Covered barley + wheat (CB + W)	136 + 246
VIII	Covered barley + covered oats + wheat (CB + CO + W)	91 + 170 + 164
IX	Naked barley + wheat (NB + W)	136 + 246
X	Covered barley + naked oats + wheat (CB + NO + W)	91 + 187 + 164

* The sowing of each component in the mixture results from the recommended quantity for each species in pure sowing in accordance with the agricultural practice.

The tillage included pre-winter ploughing. In spring, a combined implement for soil tillage was used. The seeds before sowing were treated with Scenic 080 FS (100 mL + 500 L water/100 kg grain). The row spacing was 12 cm. In the autumn, phosphorus and potassium fertilization at a dose of 20 kg ha^{-1} P_2O_5 (triple superphosphate) and 80 kg ha^{-1} K_2O (potassium salt 60%) were applied. Nitrogen fertilization in the dose of 60 kg ha^{-1} N (ammonium nitrate 34%) was applied before sowing. Sowing was performed between 5–15 April. At harvest, the grain yield of mixtures from each plot was weighed, and grain samples were taken to determine the yield sharing of individual partners in the mixture, 1000 grain weight, and total protein content. The harvest of cereal mixtures was carried out at the stage of full maturity of cereals (Biologische Bundesanstalt, Bundessortenamt i Chemical industry-BBCH 89) in the period from 6 to 15 August.

Herbicides were used during the years of research, including: 1. Puma Uniwersal 069 EW (content of the active substance: phenoxaprop-P-ethyl- 69 g L^{-1} ethylester of 2-(4-(6-chloro-1,3-benzoxazole-2-yloxy) phenoxy)propanoic acid. 2. Secateurs 125 OD (iodosulfuron-methyl-sodium 25 g L^{-1}, amidosulfuron 100 g L^{-1}), or 3. Weedlock Trio 540 SL (mecoprop (compound of the phenoxy acid group—as potassium salt)–300 g L^{-1} (24.31%) M C PA (compound of the phenoxy acid group—as potassium salt)—200 g L^{-1} (16.20%) dicamba (a compound from a group of benzoic acid derivatives—in the form of potassium salt) 40 g L^{-1} (3.24%), which effectively destroyed dicotyledonous weeds. The problem (in a few treatments) was the occurrence of wild oats on mixed plots with oats (these weeds were removed manually).

The fungicide Soligor 425 EC (active ingredient: prothioconazole 53 g L^{-1} (5.4%), spiroxamine 224 g L^{-1} (22.9%), and tebuconazole 148 g L^{-1} (15.1%), were used during the growing season. In the years of the study, there was a low level of cereal leaf beetle infestation of cereals below the economic harmfulness threshold.

The grain yield was determined at 15% humidity. Protein content was determined by the Kiejdahl method in the Main Laboratory of Chemical Analyses of IUNG-PIB in Pulawy. The grain energy value of cereal mixture was determined (taking into account the share of components) by converting the grain yield into net energy (MJ), when assuming the values calculated for pigs based on animal nutrition standards [31].

The yield suppression ratio (YSR) of the individual components of the mixture was calculated according to the methodology that Weigelt and Jolliffe presented [32]. The values of the yield suppression ratio were calculated from the ratio of the percentage share (weight) of grains of individual species in the yield to their percentage in the sowing material.

Analysis of variance (ANOVA) for randomized complete block design was performed for most data using the Statistica® software computer program package. Treatment means were compared using Tukey test at $p = 0.05$. Subsequently, six orthogonal contrast for selected treatment were performed using Statistica® software.

The following data were used to calculate the protein yield and energy value of the mixture grain yield: mixture grain yield, percentage share of components in grain yield, protein content in grain of individual components, and the value of metabolic energy of 1 kg of grain.

Methods of analyses were conducted according to methodology. Analysis of *p*-available was conducted based on the colorimetric assay and the Egner-Riehm DL method (PN-R-04023, 1996). Analysis of K-availability was conducted based on the photometric method (PN-R-04022, 1996).

3. Results

Meteorological conditions during the growing season of spring cereals (2013–2015) were not very diverse (Table 3). Average air temperatures during these growing periods were similar. The highest amount of precipitation in the growing period occurred in 2013, while the lowest occurred in 2015. This did not affect the differences in grain yields of mixtures during the years of research. The tendency for lower yields of mixtures on the better soil in 2015 can be explained by lower rainfall in that year and more permeable granulometric composition of the soil. The lack of precipitation, especially in May and June, is the main reason for low yields of cereals in a given year. It can be assumed that the sum of precipitation from March to July within the range of 220–250 mm is sufficient to obtain a fairly high spring grain yield.

Table 3. Meteorological conditions in 2013–2015 compared to the long-term (1969–2005).

Month	Total Precipitation (mm)				Daily Mean Temperature (°C)			
	2013	2014	2015	1969–2005	2013	2014	2015	1969–2005
March	19	31	29	33	−3.4	2.2	3.7	0.4
April	47	36	28	35	6.4	8.1	6.3	6.5
May	84	59	53	61	15.4	12.9	11.5	12.6
June	69	102	32	71	18	16.3	15.2	15.7
July	57	32	81	87	18.4	19.8	16.7	17.1
Total	276	260	233	283	-	-	-	-
Mean	-	-	-	-	11	11.9	10.7	10.5

Significant differences in grain yield, protein yield, and net energy yield in grain (in MJ) were found among treatments. The percentage of cereal species in the grain yield of mixtures was uneven (Tables 4 and 5). On soils of Albic Luvisols, barley exhibited a higher share in the yield (as compared to other components), which was followed by covered oats. On the Haplic Arenosols, the highest percentage in the yield was observed for covered oats, which was followed by covered barley. On the other hand, naked oats had the smallest share in grain yield of mixtures containing it on both soils.

Table 4. Yield sharing (%) of each partner in a mixture on Albic Luvisols (average 2013–2015).

Treatment	Covered Barley	Naked Barley	Covered Oats	Naked Oats	Wheat
I (CB + CO) *	55	-	45	-	-
II (CB + W)	53	-	-	-	47
III (CB + CO + W)	36	-	34	-	30
IV (NB + W)	-	50	-	-	50
V (CB + NO + W)	40	-	-	26	34
VI (CB + CO)	50	-	50	-	-
VII (CB + W)	56	-	-	-	44
VIII (CB + CO + W))	37	-	36	-	27
IX (NB + W)	-	56	-	-	44
X (CB + NO + W)	44	-	-	23	33

* Covered barley + covered oats (CB + CO), Covered barley + wheat (CB + W), Covered barley + covered oats + wheat (CB + CO + W), Naked barley + wheat (NB + W), Covered barley + naked oats + wheat (CB + NO + W), Covered barley + covered oats (CB + CO), Covered barley + wheat (CB + W), Covered barley + covered oats + wheat (CB + CO + W), Naked barley + wheat (NB + W), Covered barley + naked oats + wheat (CB + NO + W).

Table 5. Yield sharing (%) of each partner in mixture on Haplic Arenosols (average 2013–2015).

Treatment	Covered Barley	Naked Barley	Covered Oats	Naked Oats	Wheat
I (CB + CO) *	48	-	52	-	-
II (CB + W)	55	-	-	-	45
III (CB + CO + W)	34	-	39	-	27
IV (NB + W)	-	47	-	-	53
V (CB + NO + W)	36	-	-	29	35
VI (CB + CO)	46	-	54	-	-
VII (CB + W)	58	-	-	-	42
VIII (CB + CO + W)	32	-	44	-	24
IX (NB + W)	-	49	-	-	51
X (CB + NO + W)	40	-	-	28	32

* Covered barley + covered oats (CB + CO), Covered barley + wheat (CB + W), Covered barley + covered oats + wheat (CB + CO + W), Naked barley + wheat (NB + W), Covered barley + naked oats + wheat (CB + NO + W), Covered barley + covered oats (CB + CO), Covered barley + wheat (CB + W), Covered barley + covered oats + wheat (CB + CO + W), Naked barley + wheat (NB + W), Covered barley + naked oats + wheat (CB + NO + W).

The grain yield of the studied mixtures was much higher on Albic Luvisols than on Haplic Arenosols, which was conditioned by weather conditions in studied years (Tables 6–8). On both these soils, there was a large variability in grain yield between mixture variants. Regardless of the soil quality, the highest grain yields were achieved with a mixture of hulled barley and covered oats at both sowing densities. On the better soil, similarly to it, the mixture of covered barley and wheat was yielded, but only at lower sowing density. On the poorer soil, higher yields were achieved by a mixture of barley with oats and wheat regardless of sowing density. On both soils, the lowest yields were recorded for mixtures of naked grain barley with wheat and of covered barley with naked oats and wheat. All types of mixtures differing in grain species' composition were yielded similarly under both sowing densities (insignificant differences between densities) (Tables 6 and 8).

Table 6. Yields of various spring cereal mixtures on Albic Luvisols.

Treatment	Grain Yield t ha^{-1}	Protein Yield kg ha^{-1}	Net Energy Yield MJ
Yield (Y)			
2013	5.62 c	693 c	50.8 c
2014	5.89 b	726 b	53.2 b
2015	5.96 a	735 a	53.8 a
p value	<0.001	<0.001	<0.001
Cereal mixtures (M)			
I (CB + CO) **	6.30 a *	710 c	51.2 c
II (CB + W)	5.97 b	750 ab	54.6 b
III (CB + CO + W)	5.96 b	730 bc	51.2 c
IV (NB + W)	5.70 c	752 ab	57.1 a
V (CB + NO + W)	5.27 d	681 d	50.8 c
VI (CB + CO)	6.00 ab	678 d	48.4 d
VII (CB + W)	6.12 ab	760 a	55.7 ab
VIII (CB + CO + W)	5.99 b	724 b	51.2 c
IX (NB + W)	5.82 bc	768 a	56.8 a
X (CB + NO + W)	5.14 d	628 e	49.0 d
p value	<0.001	<0.001	<0.001
Y × M	<0.001	<0.001	<0.001

* Means not followed by the same letter are significantly different. ** Covered barley + covered oats (CB + CO), Covered barley + wheat (CB + W), Covered barley + covered oats + wheat (CB + CO + W), Naked barley + wheat (NB + W), Covered barley + naked oats + wheat (CB + NO + W), Covered barley + covered oats (CB + CO), Covered barley + wheat (CB + W), Covered barley + covered oats + wheat (CB + CO + W), Naked barley + wheat (NB + W), Covered barley + naked oats + wheat (CB + NO + W).

Comparison of means for grain yield, protein yield, and net energy yield by orthogonal contrasts depending on planting density proved that productivity of mixtures with covered barley depends on the sowing ratio and soil quality (Tables 7 and 9). In three mixture components, the difference of the sowing ratio of covered barley did not affect the grain yield, or yield quality. However, comparing the yields among two-component versus three-component mixtures indicates that covered barley significantly increases the productivity in higher crop density.

Table 7. Orthogonal contrast of selected treatments with covered barley depending on rate density Albic Luvisols.

Orthogonal Contrast for Tested Mixture Combinations	*p*-Value of Lineal Orthogonal Contrast		
	Seed Yield	Protein Yield	Net Energy Yield
I versus VI *	0.001 *	0.001	0.001
II versus VII	0.001	0.043	0.002
III versus VIII	n.s.	n.s.	ns
V versus X	0.001	0.001	0.001
IV versus IX	0.003	0.001	ns
III, V, VIII, X versus I, II, VI, VII	0.001	0.001	0.001

* n.s.- not significant.

Table 8. Yield of various spring cereal mixtures on Haplic Arenosols.

Treatment	Grain Yield	Protein Yield	Net Energy Yield MJ
	t ha^{-1}	kg ha^{-1}	
Yield (Y)			
2013	4.67 a	589 a	41.77 a
2014	4.70 a	593 a	42.05 a
2015	4.46 b	563 b	39.92 b
p value	<0.001	<0.001	<0.001
Cereal mixtures (M)			
I (CB + CO) ***	5.08 a *	584 ab	40.9 ab
II (CB + W)	4.65 bc	601 a	42.4 ab
III (CB + CO + W)	4.89 ab	599 a	41.6 ab
IV (NB + W)	4.41 cd	596 a	43.0 a
V (CB + NO + W)	4.14 d	550 b	41.0 ab
VI (CB + CO)	5.09 a	585 ab	40.8 ab
VII (CB + W)	4.53 c	582 ab	41.1 ab
VIII (CB + CO + W)	4.81 ab	587 a	40.4 b
IX (NB + W)	4.18 d	567 ab	40.8 ab
X (CB + NO + W)	4.36 cd	573 ab	40.6 b
p value	<0.001	<0.001	<0.001
Y × M	n.s. **	n.s.	n.s.

* Means not followed by the same letter are significantly different. ** n.s.- not significant, *** Covered barley + covered oats (CB + CO), Covered barley + wheat (CB + W), Covered barley + covered oats + wheat (CB + CO + W), Naked barley + wheat (NB + W), Covered barley + naked oats + wheat (CB + NO + W), Covered barley + covered oats (CB + CO), Covered barley + wheat (CB + W), Covered barley + covered oats + wheat (CB + CO + W), Naked barley + wheat (NB + W), Covered barley + naked oats + wheat (CB + NO + W).

Table 9. Orthogonal contrast of selected treatments with covered barley depending on rate density Haplic Arenosole.

Orthogonal Contrast for Tested Mixture Combinations	*p*-Value of Linear Orthogonal Contrast		
	Seed Yield	Protein Yield	Net Energy Yield
I versus VI	n.s.	n.s.	n.s.
II versus VII	0.049	0.030	0.035
III versus VIII	n.s.	0.162	n.s.
V versus X	0.026	0.011	n.s.
IV versus IX	0.001	0.001	0.001
III, V, VIII, X versus I, II, VI, VII	0.001	0.017	n.s.

n.s.- not significant.

Cereal species differed in terms of grain protein content covered by barley (11.4–11.9% d.m.), naked barley (12.4–12.7% d.m.), covered oats (11.1–11.5% d.m.), naked oats (13.3–13.7% d.m.), and wheat (13.7–14.4 d.m.). On poorer soil, a slightly higher protein content in grain (by 0.2–0.3% d.m.) was obtained. These data are not shown for individual components of each mixture.

On the better soil, the highest protein yields in grains were produced by mixtures of barley (covered or naked) with wheat, regardless of sowing density. Low protein yields were found in mixtures of covered barley with naked oats and wheat (especially under lower sowing density) and covered barley with covered oats under lower sowing density. On the poorer soil, higher protein yields of mixtures of barley (covered or naked) with wheat and mixtures of covered barley with covered oats and wheat were found, but only at higher sowing density (Tables 6 and 8).

The highest yield of net metabolic energy on the compared soils was recorded for a mixture of naked barley with wheat under both sowing densities as well as for a mixture of covered barley with wheat under lower sowing density. On Haplic Aerosols, the mixture of naked barley with wheat under higher sowing density gave the highest yield of net metabolic energy, while both 3-component mixtures, under lower sowing density—the lowest (Tables 6 and 8).

The yield suppression ratio of individual components of mixtures were varied (Tables 10 and 11) depending on cereal species and soil quality. The highest yield suppression ratio in the mixtures was found for covered barley, which was followed by naked barley, covered oats, and naked oats. Wheat turned out to be the least competitive in the mixture stand. On the Haplic Aerosols, covered barley and naked barley were more competitive in the mixtures (Tables 6 and 8). Oats (covered and naked) responded in the opposite way as it was more competitive on the better soil (Albic Luvisols) than on the poorer soil (Haplic Arenosols). The competitiveness of wheat was similar on both soil types.

Table 10. Yield suppression ratio of the mixture components on the soil of Albic Luvisols.

Treatment	Covered Barley	Naked Barley	Covered Oats	Naked Oats	Wheat
I (CB + CO) *	1.14	-	0.90	-	-
II (CB + W)	1.45	-	-	-	0.73
III (CB + CO + W)	1.36	-	1.11	-	0.68
IV (NB + W)	-	1.27	-	-	0.84
V (CB + NO + W)	1.33	-	-	0.94	0.83
VI (CB + CO)	1.15	-	0.89	-	-
VII (CB + W)	1.56	-	-	-	0.67
VIII (CB + CO + W)	1.28	-	1.26	-	0.60
IX (NB + W)	-	1.32	-	-	0.81
X (CB + NO + W)	1.48	-	-	0.90	0.76

* Covered barley + covered oats (CB + CO), Covered barley + wheat (CB + W), Covered barley + covered oats + wheat (CB + CO + W), Naked barley + wheat (NB + W), Covered barley + naked oats + wheat (CB + NO + W), Covered barley + covered oats (CB + CO), Covered barley + wheat (CB + W), Covered barley + covered oats + wheat (CB + CO + W), Naked barley + wheat (NB + W), Covered barley + naked oats + wheat (CB + NO + W).

Table 11. Yield suppression ratio of the mixture components on the soil of Haplic Arenosols.

Treatment	Covered Barley	Naked Barley	Covered Oats	Naked Oats	Wheat
I (CB + CO) *	1.31	-	0.78	-	-
II (CB + W)	1.39	-	-	-	0.73
III (CB + CO + W)	1.44	-	0.97	-	0.75
IV (NB + W)	-	1.35	-	-	0.79
V (CB + NO + W)	1.48	-	-	0.84	0.81
VI (CB + CO)	1.25	-	0.83	-	-
VII (CB + W)	1.51	-	-	-	0.70
VIII (CB + CO + W)	1.48	-	1.03	-	0.68
IX (NB + W)	-	1.51	-	-	0.70
X (CB + NO + W)	1.63	-	-	0.74	0.79

* Covered barley + covered oats (CB + CO), Covered barley + wheat (CB + W), Covered barley + covered oats + wheat (CB + CO + W), Naked barley + wheat (NB + W), Covered barley + naked oats + wheat (CB + NO + W), Covered barley + covered oats (CB + CO), Covered barley + wheat (CB + W), Covered barley + covered oats + wheat (CB + CO + W), Naked barley + wheat (NB + W), Covered barley + naked oats + wheat (CB + NO + W).

4. Discussion

The grain yield of the covered barley in three-component mixtures was much higher on Albic Luvisols soil than on Haplic Arenosols soil, which undermines the claim that multispecies mixtures, as an effect of specific biodiversity, grown on poorer soils, are capable of high yields. The results of the research proved that yields are determined by many interacting factors. Existing data show that equal proportion three-species mixtures may perform worse than those having a higher initial percentage of the species that is the most productive in a pure stand [3,33,34]. In our research, despite the application of half the shares of the individual components of the mixture in each combination in order to exclude the effect of species domination, we obtained the lowest yields of three-species mixtures. The yield of two-species mixtures of barley and oats was significantly better, but the yield level depended on the soil type. This is confirmed by earlier studies by Noworolnik and Terelak [35] who showed significantly higher yields of a mixture of barley and covered oats on an Albic Luvisols than on a Haplic Arenosols soil, which indicates a significant effect of habitat conditions on plant yields in the mixtures. Another aspect of the evaluation of oat-barley mixtures is the varietal selection conditioned by the structure of grain (covered vs. naked grain). Szumiło and Rachoń [36] demonstrated that higher grain yields can be obtained from barley mixtures (covered or naked) with hulled oats as compared to barley mixtures with naked oats. The above results indicate that the yields of a mixture is determined by the yielding biology of particular mixture components. This was proven by a study by Rudnicki and Wasilewska [37] who did not obtain significantly differentiated grain yields of mixtures of hulled barley with covered oats, covered barley with wheat, and barley with oats and wheat. Tobiasz-Salach et al. [38], in experiments with mixtures of covered or naked oats with other spring cereals, recorded significantly lower grain yields of hulled or naked grain mixtures of oats (hulled or naked grain) with wheat compared to the mixture of oats with covered or naked grain barley. Buczek et al. [39] showed that spring cereal mixtures (oats, barley, and wheat) yield at the level of 4.23 t ha^{-1}. On the other hand, Klima and Łabza [40] found that oats grown in mixed sowing with barley yield significantly higher than in pure sowing. In our experiment, the yield of the mixture of oat and barley sown in large sowing amounts to 6.3 t ha^{-1}.

The yield suppression ratio of individual components of mixtures were varied depending on cereal species. The highest yield suppression ratio in the mixtures were found for covered barley, which was followed by naked barley, covered oats, and naked oats. Wheat turned out to be the least competitive in the mixture stand. Presented results were partly confirmed by Klimek-Kopyra et al. [9]. The authors revealed asymmetric interspecific competition between species in two and three component mixtures. Wheat, despite having a high share in the mixture, did not display high productivity. Leszczyńska and Grabiński [41] and Czaban et al. [42] claimed that the interaction of plants in the canopy cannot be fully explained without the knowledge of allelopathy.

An important aspect that determines the suitability of plants for mixed sowings is the quality of the obtained grains. For the grain to be useful for industrial purposes, it has to exhibit a high protein content including at least 11.5% of protein in dry matter, and 14% of protein in dry matter, which is meant for improving the value of milling mixtures with low-quality grain [43]. The results of our research indicate that, due to the increased amount of protein in the compared mixtures, only the combinations with wheat are effective and appropriate for use in the fodder industry.

Mixtures of hulled barley with wheat grown on high quality soil were characterized by significantly higher protein (760 kg ha^{-1}) and metabolic energy yield (55.7 MJ). On Haplic Arenosols soil, the highest net metabolic energy yield was recorded for a mixture of naked barley with wheat at a higher sowing density, while the lowest—for both three-component mixtures at a lower sowing density. Other results were obtained by Kijora and Wróbel, [44], who proved that higher grain protein yields and net metabolic energy yields could be obtained from mixtures of covered barley with covered oats than from mixtures of barley with naked oats. Higher fat yields, however, can be obtained from mixtures of barley with naked oats.

5. Conclusions

The grain yield differentiations of cereal mixture variants was different than protein and metabolic energy yield due to the lower grain yield but higher content of protein and metabolic energy in the grains of naked forms of barley, oats, and wheat.

A higher grain yield was indicated from a mixture of cove forms of spring barley and oats. However, the yield of protein and metabolic energy of mixtures with the share of naked forms of barley, oats, and wheat was similar to that of covered forms.

Regardless of the soil quality and sowing density, the highest grain yields were obtained from a two-mixture component of covered barley with covered oats (Albic Luvisols: 6.3 t ha^{-1} and 6.0 t ha $^{-1}$, Haplic Arenosols: 5.08 t ha^{-1} and 5.09 t ha $^{-1}$, respectively). Three mixture components (CB + CO + W) lack of differentiations of cereal mixture variants in terms of yield and protein yields was noted.

Author Contributions: Conceptualization, D.L. and A.K.-K.; methodology, D.L.; software, A.K.-K.; validation, D.L., A.K.-K. and K.P.; formal analysis, A.K.-K.; investigation, D.L.; resources, D.L.; data curation, D.L.; writing—original draft preparation, A.K.-K., D.L., K.P.; writing—review and editing, A.K.-K., D.L., K.P.; visualization, A.K.-K.; supervision, D.L.; project administration, D.L.; funding acquisition, D.L. and K.P. All authors have read and agreed to the published version of the manuscript.

Funding: This research was funded by Institute of Soil Science and Plant Cultivation, State Research Institute in Puławy, Poland and was funded by University of Life Sciences in Lublin, Poland.

Acknowledgments: We would like to thank Kazimierz Noworolnik, who is an employee of the Department of Cereal Crop Production, the Institute of Soil Science and Plant Cultivation—the State Research Institute in Puławy, Poland for his contribution to the creation of scientific work and for long-term collaboration.

Conflicts of Interest: The authors declare no conflict of interest.

References

1. Altieri, M.; Nicholls, C.; Montalba, R. Technological approaches to sustainable agriculture at a crossroads: An agrotechnical prespective. *Sustainability* **2017**, *9*, 349. [CrossRef]
2. Monti, M.; Pellicano, A.; Pristeri, A.; Badagliacca, G.; Preiti, G.; Gelsomino, A. Cereal/grain legume intercropping in rotation with durum wheat in crop/livestock production systems for Mediterranean farming system. *Field Crop Res.* **2019**, *240*, 23–33. [CrossRef]
3. Klimek-Kopyra, A.; Bacior, M.; Zając, T. Biodiversity as a creator of productivity and interspecific competitiveness of winter cereal species in mixed cropping. *Ecol. Mod.* **2017**, *343*, 123–130. [CrossRef]
4. Barot, S.; Allard, V.; Cantarel, A.; Enjalbert, J.; Gauffreteau, A.; Goldringer, I.; Lata, J.-C.; Roux, X.; Niboyet, A.; Porcher, E. Designing mixtures of varieties for multifunctional agriculture with the help of ecology. A review. *Agron. Sus. Dev.* **2017**, *37*, 13. [CrossRef]
5. Godfray, H.C.J.; Beddington, J.R.; Crute, I.R.; Haddad, L.; Lawrence, D.; Muir, J.F.; Pretty, J.; Robinson, S.; Thomas, S.M.; Toulmin, S. Food security: The challenge of feeding 9 billion people. *Science* **2010**, *327*, 812–818. [CrossRef]
6. Hong, Y.; Heerink, N.; Zhao, M.; Werf, W. Intercropping contributes to a higher technical efficiency in smallholder farming: Evidence from a case study in Gaotai Country, China. *Agric. Syst.* **2019**, *173*, 317–324. [CrossRef]
7. Petersen, B.; Snapp, S.S. What is sustainale intensyfication: Views from experts. *Land Use Policy* **2015**, *46*, 1–10. [CrossRef]
8. Smith, A.; Snapp, S.; Chikowo, R.; Torne, P.; Bekunda, M.; Glover, J. Measuring sustainable intensification in smallolder agroecosystems: A review. *Glob. Food Sec.* **2017**, *12*, 127–138. [CrossRef]
9. Li, L.; Zhang, L.; Zhang, F. Crop Mixtures and the Mechanisms of Overyielding. In *Encyclopedia of Biodiversity*, 2nd ed.; Levin, S.A., Ed.; Academic Press: Waltham, MA, USA, 2013; Volume 2, pp. 382–395.
10. Malézieux, E.; Crozat, Y.; Dupraz, C.; Laurans, M.; Makowski, D.; Lafontaine, O.H. Mixing plant species in cropping systems: Concepts, tools and models: A review. *Sustain. Agric.* **2009**, *3*, 329–353.
11. Klimek-Kopyra, A.; Zając, T.; Rębilas, K. A mathematical model for the evaluation of cooperation and competition effects in intercrops. *Eur. J. Agron.* **2013**, *1*, 9–17. [CrossRef]

12. Neugschwandtner, R.W.; Kaul, H.-P. Concentrations and uptake of macronutrients by oat and pea in intercrops in response to N fertilization and sowing ratio. *Arch. Agron. Soil Sci.* **2016**. [CrossRef]

13. Lithourgidis, A.S.; Dordas, C.A.; Damalas, C.A.; Vlachostergios, D.N. Annual intercrops: An alternative pathway for sustainable agriculture. *Aust. J. Crop Sci.* **2011**, *5*, 396–410.

14. Boligłowa, E.; Gleń-Karolczyk, K.; Klimek-Kopyra, A.; Zając, T. Condition of winter wheat in pure and mixture sowing. *J. Res. Appl. Agric. Eng.* **2018**, *62*, 58–61.

15. Leszczyńska, D. Cereal mixtures–Important link in production potential of Polish agriculture. *Pam. Puł.* **2003**, *132*, 287–293. (In Polish)

16. Newton, A.C.; Guy, D.C.; Bengough, A.G.; Gordon, D.C.; McKenzie, B.M.; Sun, B.; Valentine, T.A.; Hallett, P.D. Soil tillage effects on the efficacy of cultivars and their mixtures in winter barley. *Field Crop Res.* **2012**, *128*, 91–100. [CrossRef]

17. Grettenberger, I.A.M.; Tooker, J.F. Variety mixtures of wheat influence aphid populations and attract an aphid predator. *Arthr. Plant Inter.* **2017**, *11*, 133–146. [CrossRef]

18. Zając, T.; Oleksy, A.; Stokłosa, A.; Klimek-Kopyra, A.; Styrc, N.; Mazurek, R.; Budzyński, W. Pure sowings versus mixtures of winter cereal species as an effective option for fodder grain production in temperate zone. *Field Crop Res.* **2014**, *166*, 152–161. [CrossRef]

19. Noworolnik, K.; Leszczyńska, D. Przydatność nagoziarnistego jęczmienia (Rastik) i nagoziarnistego owsa (Akt) do uprawy w zasiewach mieszanych. *Pam. Puł.* **2004**, *138*, 109–116. (In Polish)

20. Klimek-Kopyra, A.; Kulig, B.; Oleksy, A.; Zając, T. Agronomic performance of naked oat (*Avena nuda* L.) and faba bean intercropping. *Chil. J. Agric. Res.* **2015**, *75*, 168–173. [CrossRef]

21. Sterna, V.; Zute, S.; Jansone, I.; Kantane, I. Chemical composition of covered and naked spring barley varieties and their potential for food production. *Pol. J. Food Nut. Sci.* **2017**, *67*, 151–158. [CrossRef]

22. Noworolnik, K.; Leszczyńska, D. Interspecific competition in barley and oats mixtures. *Żywność. Nauka. Tech. Jakość* **1999**, *1*, 126–130.

23. Hackett, R. A comparison of husked and naked oats under Irish conditions. *Irish J. Agric. Food Res.* **2018**, *57*, 1–8. [CrossRef]

24. MacLeod, M.G.; Valentine, J.; Cowan, A.; Wade, A.; McNeill, L.; Bernard, K. Naked oats: Metabolisable energy yield from a range of varieties in broilers, cockerels and turkeys. *Br. Poult. Sci.* **2008**, *493*, 368–377. [CrossRef] [PubMed]

25. Givens, D.I.; Davies, T.W.; Laverick, R.M. Effect of variety, nitrogen fertiliser and various agronomic factors on the nutritive value of husked and naked oats grain. *Anim. Feed Sci. Technol.* **2004**, *1131*, 169–181. [CrossRef]

26. Biel, W.; Bobko, K.; Maciorowski, R. Chemical composition and nutritive value of husked and naked oats grain. *J. Cereal Sci.* **2009**, *494*, 413–418. [CrossRef]

27. Marshall, A.; Cowan, S.; Edwards, S.; Griffiths, I.; Howarth, C.; Langdon, T.; White, E. Crops that feed the world 9. Oats-a cereal crop for human and livestock feed with industrial applications. *Food Sec.* **2013**, *5*, 13–33. [CrossRef]

28. Butt, M.; Tahir-Nadeem, M.; Kashif Iqbal Khan, M.; Shabir, R.; Butt, M. Oat: Unique among the cereals. *Eur. J. Nutr.* **2008**, *47*, 68–79. [CrossRef]

29. Ločmele, I.; Legzdiņa, L.; Gaile, Z.; Kronberga, Z. Cereal variety mixtures and populations for sustainable agriculture: A Review. *Agric. Aciences. Res. Rural Dev.* **2016**, *1*, 7–14.

30. Szmigiel, A.; Oleksy, A. Yielding of naked and husked forms of spring barley and oat cultivated in mixtures and in pure stands on a soil of very good complex. *Biul. Inst. Hod. Aklim. Roś.* **2005**, *236*, 173–181.

31. AOAC. *Official Methods of Analysis: Official Method for Protein, Method No. 920.87*; Association of Official Analytical Chemists: Washington, DC, USA, 1995.

32. Weigelt, A.; Jolliffe, P. Indices of plant competition. *J. Ecol.* **2003**, *91*, 707–720. [CrossRef]

33. Buczek, J.; Tobiasz-Salach, R.; Bobrecka-Jamro, D. Ocena plonowania i odchwaszczającego działania jarych mieszanek zbożowych. *Post. Nauk Rol.* **2007**, *516*, 11–18. (In Polish)

34. Sobkowicz, P.; Tendziagolska, E.; Lejman, A. Performance of Multi-component mixtures of spring cereals. Part1. Yields and Yield compoenents. *Acta Sci. Pol. Agric.* **2016**, *15*, 25–35.

35. Noworolnik, K.; Terelak, H. Plonowanie jęczmienia jarego i owsa oraz ich mieszanki w zależności od warunków glebowych. *Rocz. Gleb.* **2005**, *3–4*, 60–66. (In Polish)

36. Szumiło, G.; Rachoń, L. Siewy czyste i mieszane nagoziarnistych ioplewionych odmian jęczmienia jarego i owsa. Czesc II. Porównanie plonowania izdrowotności. *Pos. Nauk Rol.* **2007**, *516*, 257–265. (In Polish)

37. Rudnicki, F.; Wasilewski, P. Wpływ doboru gatunków i ilości opadów na wydajność jarych mieszanek zbożowych. *Frag. Agron.* **1993**, *4*, 95–96. (In Polish)

38. Tobiasz-Salach, R.; Bobrecka-Jamro, D.; Buczek, J. Plonowanie owsa siewie czystym i w mieszankach ze zbożami jarymi. *Frag. Agron.* **2007**, *4*, 20–23. (In Polish)

39. Buczek, J.; Jarecki, W.; Bobrecka-Jamro, D. The yield and weed infestation of spring cereals cultivated in the mixture after cereal forecrops. *Prog. Plant Prot.* **2012**, *52*, 58–61.

40. Klima, K.; Łabza, T. Plonowanie i efektywność ekonomiczna uprawy owsa w siewie czystym i mieszanym w systemie ekologicznym i konwencjonalnym. *Żywność. Nauka. Tech. Jakość* **2010**, *70*, 141–147. (In Polish)

41. Leszczyńska, D.; Grabiński, J. Cereals germination in mixture stand-allelopathic aspects. *Ann. Univ. Mariae Curie-Skłodowska Agric.* **2004**, *59*, 1977–1984.

42. Czaban, J.; Wróblewska, B.; Leszczyńska, D. Alleopathic effects of barley and oats on their rhizoplane and *rhizosphere Fusarium* spp. *Alleopathy J.* **2017**, *421*, 109–122. [CrossRef]

43. Biel, W.; Maciorowski, R. Ocena wartości odżywczej ziarna wybranych odmian pszenicy. *Żywność. Nauka. Tech. Jakość* **2012**, *2*, 45–55. (In Polish)

44. Kijora, C.; Wróbel, E. Plonowanie i jakość ziarna jarych zbóż pastewnych uprawianych w siewie czystym i mieszankach. *Pam. Puł.* **2004**, *1358*, 81–90. (In Polish)

Article

Row-Intercropping Maize (*Zea mays* L.) with Biodiversity-Enhancing Flowering-Partners—Effect on Plant Growth, Silage Yield, and Composition of Harvest Material

Vanessa S. Schulz [1,2,*], Caroline Schumann [2], Sebastian Weisenburger [3], Maria Müller-Lindenlauf [1], Kerstin Stolzenburg [2] and Kurt Möller [2]

[1] Institute of Applied Agriculture, Nuertingen-Geislingen University, 72622 Nuertingen, Germany; Maria.Mueller-Lindenlauf@hfwu.de

[2] Center for Agricultural Technology Augustenberg, 76227 Karlsruhe, Germany; Caroline.Schumann@ltz.bwl.de (C.S.); Kerstin.Stolzenburg@ltz.bwl.de (K.S.); Kurt.Moeller@alumni.tum.de (K.M.)

[3] Agricultural production and control, Agricultural Office Rastatt, 76437 Rastatt, Germany; s.weisenburger@landkreis-rastatt.de

* Correspondence: vschulz89.vs@gmail.com; Tel.: +49-721-9518-216

Received: 29 September 2020; Accepted: 2 November 2020; Published: 4 November 2020

Abstract: Maize cultivation faces some challenges, particularly in terms of low biodiversity in fields. Since maize is a highly efficient and economic crop, it is cultivated on large areas in Germany, with a high share in crop rotation, especially where cattle farming takes place. Such landscapes provide less habitat and food resources for small vertebrates and arthropods. Intercropping maize with flowering partners might have a positive effect on the environment and might promote biodiversity in agricultural ecosystems. Therefore, in two-year field experiments on three sites in south-western Germany, plants were tested for their suitability as intercropping partners in maize crops (*Medicago sativa*, *Melilotus officinalis*, *Vicia sativa*, *Tropaeolum majus*, *Cucurbita pepo*, and *Phaseolus vulgaris*). Almost all tested partners produced flowers, except *M. officinalis*. Intercropping maize with *P. vulgaris* or *T. majus* achieved comparable dry matter yields as sole maize, without changes in the biomass quality. For maize-intercropping, site adapted weed control and practicable sowing technique are mandatory, which already exist for *P. vulgaris* and *T. majus*. The study shows that intercropping maize with biodiversity-enhancing flowering partners can provide an applicable alternative to sole maize cropping and enhance biodiversity. The large production areas of maize have great potential for ecological improvements in agriculture.

Keywords: maize; *Zea mays* L.; biodiversity; intercropping; silage; growth; yield; quality; legume; non-legume

---

1. Introduction

Modern agriculture has to face several challenges in the next decades. It has to secure an adequate supply of food, feed, fuel, and fiber for a growing world population under the current situation that 24 billion Mg of fertile soil are annually lost due to erosion, and an increasing land desertification [1,2]. Additionally, there is a loss of biodiversity, which is partly caused by expansion and intensification of modern agriculture [3]. In landscapes dominated by agriculture, a loss of pollinators and wildflowers as their food resources among other things can be observed [4]. While 90% of the plants are dependent on animal pollination, 58 out of 130 crop pollinating bee species in the EU are threatened. Pollinators contribute up to 35% to the global crop yields and if their numbers continue to decline, global food supply can no longer be guaranteed [5]. For their survival, pollinators need habitat diversity [6]. In most agricultural landscapes, the structural diversity declines or no longer exists [7].

To counteract these current problems, future cropping systems should be productive and sustainable in ecological, economic, and social ways. Therefore, intercropping could make a valuable contribution.

Intercropping is the practice of cultivating more than one crop simultaneously on the same area of land. There are some different forms of intercropping, like strip-, row-, mixed-, or relay-intercropping, just to mention the most popular ones [8]. The advantages of intercropping range from ecological and economic to social benefits. Gebru [9] listed the most common advantages of intercropping, which can be found in literature; it prevents soil from erosion and desiccation by more ground coverage, yields can be more stable in unsteady seasons and higher under normal growing conditions, and there is an income diversification by different crops and different working peaks.

Intercropping has always been the most widespread form of cropping system in (sub-) tropical and developing countries. Maize (*Zea mays* L.) is one of the most cultivated plants in these regions and ensures the food supply of the population. With 48.5 million Mg harvested in 2018 in the least developed countries, maize (grain and silage) ranked in second place after paddy rice, before wheat [10].

For these developing countries intercropping maize is the common practice. Maize is mostly intercropped with legumes (*Phaseolus vulgaris* L., *Glycine max* (L.) MERR., *Arachis hypogaea* L.), vegetables (*Solanum tuberosum* L., *Raphanus sativus* var. *sativus* L., *Spinacia oleracea* L., *Cucurbita* spp. L.), or cereals like wheat (*Triticum aestivum* L.) [10–18]. Mostly, legume crops were intercropped due to their ability for biological nitrogen fixation [19].

In the industrialized countries maize-intercropping is generally not used. Knörzer et al. [20] showed that in Africa and Asia the small farm structure and the land scarcity make intercropping common.

Especially the cropping of maize related to some negative environmental effects due to its growth habits and cropping system. Maize has a slow initial development and late canopy closure, which encourage erosion and nitrate leaching into groundwater bodies [20–24].

Maize cultivation covers large areas of land. In Germany, 2.6 million ha out of 11.7 million ha of arable land were cropped with maize in 2018, 84% for silage maize production [25]. In the federal state of Baden-Württemberg, 17% of the 814,600 ha arable land were cultivated with silage maize and 7% with maize for grain use in the same year [26].

Therefore, intercropping could be an option to increase the biodiversity in agricultural ecosystems [27]. The additional cropped plants will elongate the flowering period which could offer a food basis and create habitats for small vertebrates and some arthropods (hymenoptera, coleoptera, lepidoptera, and diptera). Additional ground coverage by the partner can reduce nitrate leaching and erosion. A study from Denmark with silage maize showed that intercropping with *Festuca rubra* (L.) could reduce nitrate leaching by 15–37% (depending on soil type and crop rotation) [28]. Yields comparable to those in monocropping are achievable, as shown in Iran. Javanmard et al. [29] showed that intercropping of maize with vetch (*Vicia villosa* ROTH), bitter vetch (*Vicia ervilia* (L.) WILLD.), berseem clover (*Trifolium alexandrinum* L.), and common bean (*Phaseolus vulgaris* L.) resulted in higher biomass yields due to the use of different soil layers by the root systems of the different intercropping partners By increasing the crude protein of the harvest material, the purchase of protein feed might be reduced. Especially for the regions in Baden-Württemberg, such as the Markgräfler Land, Bruchsal-Mannheim-Heidelberg, Kraichgau, Stuttgart-Heilbronn, Main-Tauber-Kreis, and Upper Swabia with a high biogas plant and/or livestock density and maize cultivation in general and also high nitrate loads in the groundwater, maize intercropping could be an interesting alternative [30].

Therefore, if maize cultivation uses large areas of land, and the current cultivation system has negative effects on the environment (monoculture, late canopy closing, nitrate leaching, limited habitats, and food resources), and small vertebrates and arthropods, then intercropping partners that provide an additional flowering aspect and ground coverage can be beneficial for biodiversity. To the best of our knowledge, there are only few approaches to create a biodiversity aspect in silage maize with *Phaseolus* beans or flowering mixtures [31–35]. Therefore, the objectives of this study were to: (i) test different legume and non-legume plants for their suitability for intercropping systems with maize, and to enhance the flowering aspect in silage maize stands; (ii) determine the effects of these partners on maize growth and yield; and (iii) determine the effects on composition of the harvested biomass.

2. Materials and Methods

2.1. Site Conditions and Climate

The field experiments were carried out from 2018 to 2019 on two experimental sites at the Centre for Agricultural Technology Augustenberg (Ettlingen and Forchheim am Kaiserstuhl) and at the experimental station Tachenhausen of the Nuertingen-Geislingen University in Southwest Germany (Table 1).

Table 1. Site characteristics of the three experimental sites Ettlingen (ET), Tachenhausen (TH), and Forchheim am Kaiserstuhl (FAK).

	Ettlingen	Tachenhausen	Forchheim am Kaiserstuhl
	(ET)	(TH)	(FAK)
Coordinates	N 48°56′ E 8°23′	N 48°39′ E 9°23′	N 48°10′ E 7°41′
Cropping System	conventional	conventional	organic
Elevation above sea level	135 m	360 m	175 m
Σmean long-term annual precipitation	742 mm	802 mm	882 mm
Average long-term air temperature	10.1 °C	10.2 °C	10.2 °C
Edaphoclimatic area	Rhine valley and site valleys	Upper Gaeu region	Rhine valley and site valleys
Geology	Loess sediment	Loess sediment	Loess sediment on Würm gravel
Soil type	Luvisol	Luvisol	Luvisol
Soil texture	sandy/silty loam	clay loam	sandy loam

The weather in 2018 was dry and hot, with only 261 mm precipitation during the main growing season (March to October) at ET (deficit of 270 mm, compared to the mean long-term annual precipitation), 388 mm at TH (deficit of 153 mm) and 290 mm at FAK (deficit 371 mm). Spring 2019 was more favorable than in 2018. April was rather warm, while May was cold. Summer was warm again. During the main growing season in 2019 482 mm precipitation occurred in ET (deficit 49 mm), 597 mm in TH (deficit 76 mm), while with 585 mm in FAK, a precipitation plus of 56 mm was documented. Climate charts can be found in the Appendix A (Figures A1–A3).

2.2. Experimental Design

At ET, the previous crops were sweet corn (*Zea mays* L.) in summer 2017 and a mixture of *Sinapsis alba* (L.) and *Raphanus sativus* var. *oleiformis* (Pers.) as green manure over winter. In TH 2017, winter wheat (*Triticum aestivum* L.) and in 2018 spring barley (*Hordeum vulgare* L.) were grown as previous crops followed by a fallow over both winter times. In FAK in both years, the previous crop was potato (*Solanum tuberosum* L.), also followed by fallow over winter.

ET was only rated in 2018. Therefore, a split-split-plot design with three replicates was used. The three factors tested were; the amount of nitrogen fertilizer (N-Level, main plot), the placement of the IFP seeds (seed placement, subplot 1) and nine different intercropped flowering partners (2018) (IFP, subplot 2). The three levels of nitrogen fertilization consist of 0%, 50%, and 100% of the required nitrogen demand of a sole silage maize crop. The seed placement of the IFP was either between the maize rows (BR) or in the maize rows (IR). Therefore, the IFP was sown close to the maize rows, which should simulate a simultaneous sowing of maize and IFP. Sowing rates of the IFP were set according to the amount used to establish a sole crop stand (Table 2). The flowering partners for intercropping were chosen according to their flowering properties and the attractiveness/food supply for insects. Nitrogen fertilization took place on 25 April 2018 and 19 April 2019. The required fertilizer nitrogen amount in 2018 was 172 kg ha^{-1} under 100% and 86 kg ha^{-1} under 50%. The used fertilizer was ALZON 46 (46% total-nitrogen as urea with 2-cyanoguanidine and 1,2,4-triazole as nitrification inhibitor). Fertilizer amount was based on accepted local fertilization recommendations, taking residual soil nutrient from soil tests into account.

Table 2. The used intercropped flowering partners (IFPs), varieties, sowing rates, and the 1000-grain mass.

Treatment	IFP	Variety	Sowing Rate	1000-Grain Mass (g)
M Maize (Control)		Figaro	8 seeds m^{-2}	245
MA Maize +	Alfalfa (*Medicago sativa* L.)	Catera	15 kg ha^{-1}	2.5
MC Maize +	Yellow sweet clover (*Melilotus officinalis* L.)	-	4 kg ha^{-1}	2.0
MV Maize +	Common vetch (*Vicia sativa* L.)	Jose [#] Mery [¶]	70 kg ha^{-1}	40–60
MN Maize +	Nasturtium (*Tropaeolum majus* L.)	-	20 kg ha^{-1}	200
MS1 Maize +	Summer squash I (*Cucurbita pepo* L.)	Jack be little	1.6 seeds m^{-2}	55
MS2 Maize +	Summer squash II (*Cucurbita pepo* L.)	Spinnig/Dancing Gourd	1.6 seeds m^{-2}	27
MB1 Maize +	Common bean I (*Phaseolus vulgaris* L.)	WAV 512/612 [†]	4.5 seeds m^{-2}	190
MB2 Maize +	Common bean II (*Phaseolus vulgaris* L.)	Anellino verde [‡]	4.5 seeds m^{-2}	350
MS3 Maize +	Summer squash III [‖] (*Cucurbita pepo* L.)	New England Pie	1.6 seeds m^{-2}	102
MM1 Maize +	Mixture I [‖] (Summer squash II + common bean I)		1.6 seeds m^{-2} + 4.5 seeds m^{-2}	
MM2 Maize +	Mixture II [‖] (Common vetch + common bean I)		70 kg ha^{-1} + 4.5 seeds m^{-2}	

[#] conventional variety; [¶] organic variety; [†] WAV 512 was renamed in 2018 to WAV 612; [‡] Anellino verde not available in 2019; [‖] Only tested in 2019.

At TH and FAK, a randomized complete block design with four replicates, and nine (2018) and 12 (2019) IFP treatments were used, respectively. In TH and FAK, no nitrogen was applied (0% N), and the IFP were only sown in the maize rows (IR).

Maize and IFP were sown on the same day (ET 3 May 2018, TH 28 May 2018 and 15 May 2019, FAK 9 May 2018 and 14 May 2019). Maize (*Zea mays* L., cv. "Figaro") was sown at 0.03 m sowing depth, with a row distance of 0.75 m and a planting density of 8 plants m^{-2} by a four-row pneumatic precision planter. Each plot consisted of four rows of maize. Four rows of IFP were sown after maize sowing by a plot seeder (type "Hege 80 PNI," Zürn Harvesting GmbH & Co. KG. Schöntal-Westernhausen, Germany, sowing width = 3 m, 4 rows with 0.75 m inter-row spacing). The sowing rate of 8 maize plants m^{-2} was chosen to enable the comparison to the maize–common bean intercropping treatments. In this system, maize is typically established with 8 plants m^{-2}.

At the conventional managed location ET in 2018 weed control was done in treatment M, MS1, MS2, MB1 and MB2 with a mixture of 2.8 L ha^{-1} *Pendimethalin* (455 g active ingredient L^{-1}) and 1.0 L ha^{-1} *Dimethenamid-P* (720 g a.i. L^{-1}) as pre-emergence treatment. Treatment MN was treated in pre-emergence with 1.75 L ha^{-1} *Pendimethalin* (455 g a.i. L^{-1}). Treatment MA and MC were treated in post-emergence with 2.2 L ha^{-1} *Pendimethalin* (455 g a.i. L^{-1}). At TH in 2018 only treatment M, MB1 and MB2 were treated with 2.8 L ha^{-1} *Pendimethalin* (455 g a.i L^{-1}) and 1.0 L ha^{-1} *Dimethenamid-P* (720 g a.i. L^{-1}) as pre-emergence treatment and treatment MA was treated with 2.2 L ha^{-1} *Pendimethalin* (455 g a.i. L^{-1}) in post-emergence. In 2019, herbicide application for M, MB1, MB2, and MA were the same as in 2018, additionally MV was treated with 2.2 L ha^{-1} *Pendimethalin* (455 g a.i L^{-1}) in post-emergence and treatment MN with 1.75 L ha^{-1} *Pendimethalin* (455 g a.i. L^{-1}) in pre-emergence. No chemical weed control was done in MS1, MS2, MS3, MM1, and MM2. Under organic management hoeing was carried out in 2018 three times (26 May 2018, 28 May 2018, and 5 June 2018). In 2019, the experimental site was two times cultivated by tine harrow (24 May 2019 and 27 May 2019) and three times hoed (1 June 2019, 13 June 2019, and 18 June 2019). In TH and FAK, hand weeding was done on demand. Applications of *Trichogramma brassicae* (Latreille) against *Ostrinia nubilis* (Hübner) were done according to the recommendations [36].

Plots were 12 m (ET), 11 m (TH), 10 m (FAK) long, and 3 m broad, consisting of a total of four rows maize and four rows of IFP per plot. Core plots for chopping were 10 m (ET), 9 m (TH), 8 m (FAK) long, and 1.5 m broad including two maize and IFP rows and leaving two rows of maize and IFP on the left and the right of the plot as a border. The complete experiments were enclosed to the right and the left by border plots (four maize rows, 0.75 cm row spacing) in order to protect the experiments from external influences.

Harvest was done by a plot harvester (type "BAURAL SF 2000", Zürn Harvesting GmbH & Co. KG, Schöntal-Westernhausen, Germany, cutting width = 1.50 m) at dough stages of maize (ET 13 August 2018, TH 31 August 2018 and 17 September 2019, FAK 24 August 2018 and 5 September 2019).

2.3. Data Collection and Statistical Analysis

The dates of beginning and end of flowering of maize and the IFP were determined. For maize flowering was the time of anthesis (♂: from tassel visible until flowering completed; ♀: tip of ear emerging from leaf sheath until stigmata completely dry; [37]), for IFP when the first flowers were present until no more flowers were present. Prior to harvest plant height from soil surface to tassel tip of maize was measured. To determine the share of maize and IFP in the harvested biomass in ET a section harvest of 1.5 m × 0.67 m, in TH 1.5 m × 1.0 m and in FAK 1.5 m × 0.33 m was cut two weeks before harvest. Cutting height was the height at which the plot harvester later cuts the crops (0.15 m). All plant biomass in this area (maize, IFP, and weeds) was cut. The total weights of the fractions were determined, and the dry matter yield (DMY) and dry matter content (DMC) were determined gravimetrically after oven drying the material at 105 °C for 48 h.

Due to the high number of plots in ET in 2018 only 0% and 100% nitrogen fertilizer levels and only IR sowing were rated in combination with the IFP. Common vetch (MV) was not included due to a high weed infestation. In plots with the summer squashes I and II (MS1, MS2) no squash grew up, and were excluded from statistical analysis. In FAK treatment MA in 2018, and in 2019, the treatments MA, MC, MS1, MS2, and MM1 were not tested. The alfalfa in treatment MA was damaged due to mechanical weed management, also the yellow sweet clover (MC). Both squashes (MS1, MS2) had no biomass contribution to the yield due to the flat growth habits. The field emergence of the squashes in mixture I (MM1) was uneven with large gaps. Such strong weed infestation as seen in ET did not occur in TH and FAK. In addition, weeds that had survived the chemical and mechanical weed treatments in TH and FAK could be controlled by hand weeding due to the smaller scope of the test compared to ET. The sites in TH and FAK also did not had as much weed potential as ET.

At harvest all plants of all core plots were chopped by plot harvester (0.15 m above surface level). The chopper gravimetrically determined the fresh weight of the biomass. Two samples of 2 kg of each plot were extracted; one sample was dried for 48 h at 105 °C to determine the DMY and DMC, and the other sample was dried for 48 h at 60 °C for chemical analysis.

Chemical analyses of the plant material were done for the parameter's crude protein (CP), crude fat (CL), crude fiber (CX), and crude ash (CA). The analyses were done as described in Bassler [38]. The nitrogen-free extracts (NfE) was calculated as the difference between 100% and the sum of the percentage amount of CP, CF, CX, and CA. Biogas and methane yield were calculated after the formula of Schattauer and Weiland [39]. The calculation of the feed parameters gross energy (GE), metabolizable energy (ME) and net energy for lactation (NEL) for dairy cattle feeding were done after Steinhöfel et al. [40]. Soil mineral N (NO$_3$-N) analyses "after harvest" (direct after chopping) and at "end of vegetation period" (= end of growing season, when the weather causes a plant growth stop) were done separately for the three soil layers, 0–30 cm, 30–60 cm, and 60–90 cm. After drying (105 °C), milling (<2 mm), and homogenization, 100 mL of $CaCl_2$ were added to 25.0 g of soil material. After overhead shaking (30 min, 30 rpm), the suspension was filtered and the NO_3-N was determined by continuous-flow analysis [41,42].

Statistical analyses were done with the free software R (version 3.6.2) as mixed-model. After finding significant differences via F-Test, differences between treatments were compared at α = 5% using Tukey's Honestly Significant Difference (HSD) test. For creating the letter, the display package "multcomp" was used [43].

The experimental design was a randomized complete block design with four replicates for TH and FAK, where only IFP treatments were tested. For analyzing the growth, yield and quality parameters the used, fitted model was

$$y_{ij} = \mu + r_i + g_j + e_{ij} \tag{1}$$

where y_{ij} is the response, μ the general effect, r_i the fixed effect of the *i*-th replicate, g_j the fixed effect of the *j*-th IFP, and e_{ij} the residual error of y_{ij}.

In ET, the experimental design was a three-times replicated split-split-plot. The main plots were the level of nitrogen fertilization (0%, 50%, and 100%). The subplots 1 were the position of the IFP (BR vs. IR). Subplots 2 were the different IFP treatments. For analyzing the growth, yield, and quality the used, fitted model was

$$y_{ijkl} = \mu + r_i + d_j + o_k + g_l + (rd)_{ij} + (rdo)_{ijk} + (dg)_{jl} + (og)_{kl} + (dog)_{ikl} + e_{ijkl} \tag{2}$$

where y_{ijkl} is the response, μ the general effect, r_i the fixed effect of the *i*-th replicate, d_j the fixed effect of the *j*-th nitrogen level, o_k the fixed effect of the *k*-th seed placement, g_l the fixed effect of the *l*-th IFP, rd_{ij} the random interaction between the *i*-th replicate and the *j*-th nitrogen level, rdo_{ijk} the random interaction between the *i*-th replicate, the *j*-th nitrogen level and the *k*-th seed placement, dg_{jl} the random interaction between the *j*-th nitrogen level and the *l*-th IFP, og_{kl} the random interaction between the *k*-th seed placement and the *l*-th IFP, dog_{ikl} the random interaction between the *i*-th nitrogen level, the *k*-th seed placement, and the *l*-th IFP and e_{ijkl} is the residual error of y_{ijkl}.

Normal distribution and homogeneity of variance were checked graphically.

Due to the weather extreme the statistical analyses were done for each year separate; the significant effect of the year masked the variance within the IFP. The analyses were done separately for each experimental location due to the differences in management and the differences in experimental set-up (ET: conventional, three-factorial, TH: conventional, one-factorial and FAK: organic, one-factorial). Pearson's coefficient of correlation for ET in 2018 were calculated using the R package "corrplot" [44].

3. Results

3.1. Flowering Period, Plant Growth, and Soil Nitrate Content

Experiments showed that maize flowering started between 60 and 70 days after sowing (DAS) and lasted for a short period between nine (FAK 2018) and 21 days (ET 2018) (Table 3). Beside of yellow sweet clover (MC), all IFP provided an additional period with flowers. Alfalfa (MA) was buried during the mechanical weeding process in FAK. Common vetch (MV) had the earliest flowering initiation of all IFP at all locations and years. On average, 40 DAS flowering started. However, flowering did not last for a long time in 2018 in ET and TH. After nine and 28 days of flowering, the common vetch was infested by *Erysiphaceae* (TUL. & C. TUL.) spp. and the plants died. Since the flowers of summer squashes I–III (MS1, MS2, MS3) and nasturtium (MN) were below the cutting height of the crop chopper, these IFP build up flowers even after maize harvest. In case of the common beans (MB1, MB2), it was observed that the first flowers appeared very late, more than 60 DAS. They even flower after maize anthesis and sometimes shortly before harvest.

Table 3. Days after sowing (DAS) until begin of flowering (SBF) and end of flowering (SEF), and the whole flowering time of maize and the different intercropped flowering partner (IFP) treatments (Δ = flowering period in days) for Ettlingen (ET), Tachenhausen (TH), and Forchheim am Kaiserstuhl (FAK) in 2018 and TH and FAK in 2019. The flowering periods of maize in the intercropping treatments corresponded to those of the maize in the control. For reasons of simplicity, the intercropped treatments only show the flowering periods of the IFP.

	2018									2019					
	ET			TH			FAK			TH			FAK		
	SBF	**SEF**	Δ	**SBF**	**SEF**	Δ	**SBF**	**SEF**	Δ	**SBF**	**SEF**	Δ	**SBF**	**SEF**	Δ
IFP	(DAS)		(Days)	(DAS)		(Days)	(DAS)		(Days)	(DAS)		(Days)	(DAS)		(Days)
M	65	86	21	58	72	14	63	72	9	71	79	8	70	88	18
MA	54	74	20	66	86	20	-	-	-	54	92	38	-	-	-
MC	-	-	-	-	-	-	-	-	-	-	-	-	-	-	-
MV	42	51	9	38	66	28	45	76	31	41	79	38	42	57	15
MN	45	88	43	46	86	40	46	‡		48	‡		52	‡	
MS1	-	-	-	49	86	37	65	‡		54	‡		59	‡	
MS2	-	-	-	49	86	37	56	‡		54	‡		62	‡	
MB1	74	84	10	66	86	20	76	107	31	79	125	46	95	114	19
MB2	63	84	21	66	86	20	68	107	39	-	-	-	-	-	-
MS3										48	‡		50	‡	
¶ **MM1**										79	‡		62	‡	
¶ **MM2**										79	‡		42	114	72

Common bean II (MB2) not available in 2019. M Maize (Control), MA Maize + Alfalfa, MC Maize + Yellow sweet clover, MV Maize + Vetch, MN Maize + Nasturtium, MS1 Maize + Summer Squash I, MS2 Maize + Summer Squash II, MB1 Maize + Common Bean I, MB2 Maize + Common Bean II, MS3 Maize + Summer Squash III, MM1 Maize + Mixture I, MM2 Maize + Mixture II. ¶ In mixtures the first flowers in the crop stand were counted as begin flowering, regardless which of the mixture partners flowered, and for end of flowering, vice versa ‡ IFP still flowers after harvest. If common bean flowered until harvest, harvest date was set as end of flowering due to the fact that all flowers are in the harvested material.

The combined cultivation of maize and IFP did not affected the final height of maize in ET 2018 (Table 4). At ET no significant influence of N-level ($p = 0.055$) or intercropped seed placement ($p = 0.326$) on final maize plant height were observed in 2018 (Table 4). The IFP also did not show differences on plant height ($p = 0.817$). At TH ($p = 0.943$) and FAK ($p = 0.225$), there were also no significant changes in plant height in 2018 (Tables 5 and 6). In 2019, TH ($p < 0.001$) and FAK ($p < 0.001$) had significant differences in plant height. At TH, the highest values for maize plant height were found in the control, without any intercropping partner (M) with 341 cm, while the height of the maize in the intercropping plots with nasturtium (MN), summer squash I and II (MS1, MS2), and mixture I (MM1) was not significantly different from the control (M). In FAK, only common vetch (MV) and mixture II (MM2) were significantly different from the control (M), with 253 cm and 259 cm compared to 291 cm (Table 6).

In 2018, the DMY was significantly influenced by N fertilization in ET ($p = 0.004$) (Table 4). Full fertilizer application yielded the highest DMY, while there was no difference whether 100% of the required N is applied or 50% (Table 4). In 2018, no differences were measured whether the IFP was sown in the rows or between. The strongest influence on DMY was caused by the IFP used ($p < 0.001$). Significant lower yields compared to the control (M) were found by intercropping with alfalfa (MA), yellow sweet clover (MC) and common vetch (MV). Intercropping with nasturtium (MN) or common beans (MB1, MB2) showed no significant difference compared to 15.6 Mg ha^{-1} DMY in the control. At TH significant IFP influences on DMY in both years were determined ($p = 0.003$ and <0.001) (Table 5). In 2018, common vetch (MV) showed with 11.5 Mg ha^{-1} the lowest DMY, which were not different from the control (M) in 2018, while in 2019 common vetch (MV) and mixture II (MM2) had significant lower DMY compared to the control (M), with 17.9 Mg ha^{-1}. Significant differences in the DMY were only found in 2019, not in 2018 (Table 6). Significantly lower DMY were found for common vetch (MV), summer squash III (MS3), and mixture II (MM2) in 2019.

The fractioning of the plant biomass in ET 2018 showed no significant influence of the N-Level on the share of maize, IFP and weed ($p = 0.504$, $p = 0.067$ and $p = 0.198$) (Table 4). The use of an IFP did not reduce the share of maize ($p = 0.341$). The IFP with the highest share in the harvested biomass were the common beans (MB1, MB2). They had a share of 3.45 and 4.28%, while nasturtium (MN) with 0.61% had a low share of the harvested biomass. With a share of more than 87.0% maize was the main yield component. At TH in 2018 a significant influence by IFP on the share of maize biomass was observed ($p < 0.001$) (Table 5). The smallest proportions of maize were found with 94.5% when intercropped with common vetch (MV). In 2019 there was also a significant influence of the IFP on the maize proportion ($p < 0.001$). Smallest maize proportion was found under intercropping with both mixtures (MM1, MM2) and with common bean I (MB1). FAK in 2018 showed no significant influence on maize proportion by IFP ($p = 0.221$) (Table 6). In this year also there was a weed share between 1.55% (MS2) and 6.67% (MB2). The highest proportion of IFP in the harvested biomass was found in common vetch (MV) in 2018 and mixture II (MM2) in 2019 at 5.78% and 3.96%, respectively. Both years and all locations showed that intercropping maize leads to maize proportions of >81.8% in the harvested biomass.

Table 4. Growth parameters plant height (cm), dry matter yield (DMY, Mg ha^{-1}), dry matter content (DMC, %), and the share of maize, intercropped flowering partners (IFP), and weed of the DMY from section harvest (%); for the three factor levels N-level, seed placement of IFP and the different IFP treatments at Ettlingen (ET) in 2018.

						Share at DMY of		
	Plant Height	**DMY**		**DMC**		**Maize**	**IFP**	**Weed**
	(cm)	**(Mg ha^{-1})**		**(%)**			**(%)**	
				N-Level				
0%	253	13.2	a	32.3		90.2	2.56	10.1
50%	258	14.2	ab	32.7				
100%	249	14.5	b	32.2		92.0	1.80	5.44
p-value	0.055	0.004 **		0.536		0.504	0.067	0.198
				Seed placement				
BR	254	14.2	a	35.5				
IR	253	13.7	a	32.2				
p-value	0.326	0.004 **		0.294				
				IFP				
M	244	15.6	c	32.6	ab			
MA	258	13.4	b	32.7	ab	87.0	0.91 a	9.61
MC	245	13.6	b	32.7	ab	89.6	1.26 a	10.8
MV	250	10.5	a	31.0	a			
MN	261	14.2	bc	34.0	b	92.5	0.61 a	5.04
MS1	-	-		-		-		
MS2	-	-		-		-		
MB1	261	15.5	c	32.4	ab	94.7	3.45 b	4.63
MB2	254	15.1	c	31.4	a	92.2	4.28 b	8.78
p-value	0.817	<0.001 ***		0.012 *		0.341	<0.001 ***	0.521

Values with the same letter within one parameter indicate non-significant differences within the three factor levels (N-Level, Seed placement and IFP) (HSD-test, $\alpha = 5\%$). *** $p \leq 0.001$, ** $p \leq 0.01$, * $p \leq 0.05$. M Maize (Control), MA Maize + Alfalfa, MC Maize + Yellow sweet clover, MV Maize + Vetch, MN Maize + Nasturtium, MS1 Maize + Summer Squash I, MS2 Maize + Summer Squash II, MB1 Maize + Common Bean I, MB2 Maize + Common Bean II, MS3 Maize + Summer Squash III, MM1 Maize + Mixture I, MM2 Maize + Mixture II.

Table 5. Growth parameters plant height (cm), dry matter yield (DMY, Mg ha^{-1}), dry matter content (DMC, %), and the share of maize and intercropped flowering partners (IFP) of the DMY from section harvest (%); for the different IFP treatments at Tachenhausen (TH) 2018 and 2019.

							Share at DMY of			
	Plant Height		DMY		DMC		Maize		IFP	
	(cm)		(Mg ha^{-1})		(%)		(%)			
2018										
M	303		13.4	ab	28.7	ab	100	e		
MA	299		13.0	ab	32.3	bc	97.3	bd	2.66	ab
MC	302		14.5	b	33.7	c	97.8	cd	2.21	ab
MV	297		11.5	a	30.6	ac	94.5	a	5.51	c
MN	303		12.9	ab	31.1	ac	98.5	de	1.51	a
MS1	299		12.6	ab	29.0	ac	96.0	abc	3.98	bc
MS2	307		13.4	ab	30.5	ac	96.0	abc	4.48	bc
MB1	301		12.2	a	27.0	a	95.1	ab	4.42	bc
MB2	301		13.1	ab	29.0	ac	94.9	a	5.13	c
p-value	0.943		0.003 **		0.003 **		<0.001 ***		<0.001 ***	
2019										
M	341	d	17.9	bc	32.7	a	100	d		
MA	299	ab	15.7	ab	32.0	a	96.9	cd	3.13	a
MC	305	abc	16.5	bc	32.4	a	94.5	bc	5.45	a
MV	280	a	14.2	a	30.6	a	93.6	bc	8.64	a
MN	337	cd	18.3	c	32.7	a	97.6	cd	3.42	a
MS1	337	d	17.7	bc	32.5	a	96.6	cd	3.38	a
MS2	337	d	17.6	bc	32.2	a	97.4	cd	3.76	a
MB1	334	cd	18.6	c	29.5	a	83.2	a	16.6	c
MB2	-		-		-		-		-	
MS3	294	ab	15.8	ab	32.1	a	90.6	b	9.43	ab
MM1	324	bd	18.7	c	30.0	a	81.8	a	18.0	c
MM2	286	a	14.1	a	29.1	a	84.8	a	15.2	bc
p-value	<0.001 ***		<0.001 ***		0.010 *		<0.001 ***		<0.001 ***	

Values with the same letter within one parameter indicate non-significant differences within the IFP (HSD-test, $\alpha = 5\%$). *** $p \leq 0.001$, ** $p \leq 0.01$, * $p \leq 0.05$. Common bean II (MB2) not available in 2019. M Maize (Control), MA Maize + Alfalfa, MC Maize Yellow sweet clover, MV Maize + Vetch, MN Maize + Nasturtium, MS1 Maize + Summer Squash I, MS2 Maize + Summer Squash II, MB1 Maize + Common Bean I, MB2 Maize + Common Bean II, MS3 Maize + Summer Squash III, MM1 Maize + Mixture I, MM2 Maize + Mixture II.

In 2018, in ET, the soil NO$_3$-N content was mainly influenced by the N-level ($p < 0.001$), in 0–30 cm and 30–60 cm depth after harvest, and at end of vegetation period, respectively. IFP showed in ET no influence after harvest (Figure 1). The seed placement also did not show an influence at any layer and both dates.

At the end of the vegetation period, the layers from 0–30 cm and 30–60 cm showed significant differences by the IFP used ($p = 0.044$ and $p = 0.014$) (Figure 1). Common bean II (MB2) had the significant highest amount with 80.4 kg ha^{-1} NO$_3$-N in the upper layer, while in the middle layer both beans (MB1, MB2) did not differ significantly from the control (M). In TH, no significant influence by IFP on soil NO$_3$-N content was found in any soil layers, for the two dates and in 2018. In 2019, mixture II (MM2) had significant lower NO$_3$-N contents with 3.14 kg ha^{-1} compared to 8.41 kg ha^{-1} NO$_3$-N in the control (M) at the end of the vegetation period in 30–60 cm. Also, at the 60–90 cm layer common vetch (MV) had higher NO$_3$-N contents than control (M), 3.25 compared to 1.49 kg ha^{-1} NO$_3$-N.

Table 6. Growth parameters plant height (cm), dry matter yield (DMY, Mg ha^{-1}), dry matter content (DMC, %) and the proportion of maize, intercropped flowering partners (IFP) and weed of the DMY from section harvest (%); for the different IFP treatments at Forchheim am Kaiserstuhl (FAK) 2018 and 2019.

					Share at DMY of	
	Plant Height	DMY	DMC	Maize	IFP	Weed
	(cm)	(Mg ha^{-1})	(%)		(%)	
2018						
M	234	10.5 [a]	39.7	96.1		3.94
MA	239	10.8 [a]	38.1			
MC	246	11.8 [a]	40.6	95.6	1.37 [a]	3.07
MV	234	10.7 [a]	38.3	90.5	5.78 [b]	4.90
MN	245	9.89 [a]	39.2	93.5	1.07 [a]	5.46
MS1	250	12.0 [a]	40.7	91.1	4.23 [ab]	4.38
MS2	249	12.1 [a]	40.3	94.8	2.30 [ab]	1.55
MB1	237	10.2 [a]	37.9	94.2	2.28 [ab]	4.71
MB2	240	11.0 [a]	39.3	91.9	0.44 [a]	6.67
p-value	0.225	0.032 *	0.096	0.221	0.003 **	0.310
2019						
M	291 [bc]	16.9 [d]	34.6 [bc]			
MA	294 [c]	15.7 [cd]	34.1 [ac]			
MC	296 [c]	16.8 [d]	35.2 [c]			
MV	253 [a]	12.9 [ab]	35.3 [c]	99.2 [ab]	1.16 [a]	
MN	290 [bc]	15.2 [bcd]	34.6 [bc]	99.7 [b]	0.26 [a]	
MS1	294 [c]	16.1 [cd]	34.6 [bc]			
MS2	296 [c]	16.8 [d]	34.0 [ac]			
MB1	290 [bc]	15.7 [cd]	33.1 [ac]	96.4 [a]	3.71 [b]	
MB2	-	-	-			
MS3	279 [b]	14.2 [ac]	34.9 [c]	99.1 [ab]	0.92 [a]	
MM1	297 [c]	15.7 [cd]	31.6 [ac]			
MM2	259 [a]	12.3 [a]	32.0 [ab]	98.0 [ab]	3.96 [b]	
p-value	<0.001 ***	<0.001 ***	<0.001 ***	0.021 *	<0.001 ***	

Values with the same letter within one parameter indicate non-significant differences within the IFP (HSD-test, α = 5%). *** $p \leq 0.001$, ** $p \leq 0.01$, * $p \leq 0.05$. Common bean II (MB2) not available in 2019. M Maize (Control), MA Maize + Alfalfa, MC Maize + Yellow sweet clover, MV Maize + Vetch, MN Maize + Nasturtium, MS1 Maize + Summer Squash I, MS2 Maize + Summer Squash II, MB1 Maize + Common Bean I, MB2 Maize + Common Bean II, MS3 Maize + Summer Squash III, MM1 Maize + Mixture I, MM2 Maize + Mixture II.

Differences at FAK after harvest in 2018 were only detectable in the deepest layer. Common vetch (MV) and common bean II (MB2) showed increased contents of 103% and 125% compared to the control (M), respectively. At the end of the vegetation period, intercropping with yellow sweet clover (MC) showed NO$_3$-N contents of 17.1 kg ha^{-1}, while the control had 9.48 kg ha^{-1} in the upper layer.

3.2. Quality Parameters

In 2018, the DMC was neither significantly influenced by N-level (p = 0.536), nor by seed placement of the IFP (p = 0.294) (Table 4). Only the IFP itself had a significant influence on the DMC (p = 0.012). With 34.0% nasturtium intercropping (MN) showed the highest DMC, which did not differ from the control. All DMC were in a range were ensiling could take place. At FAK (Table 6), no differences were found in 2018 (p = 0.096), but the DMC for some IFP were over the recommended ensiling maximum of 40% DMC due to the hot and dry weather, which accelerated the ripening. For TH in 2018, significant differences were observed (p = 0.003), DMC of yellow sweet clover (MC) intercropped plots were significantly higher from the control (M). In 2019, no significant differences were found between the IFP.

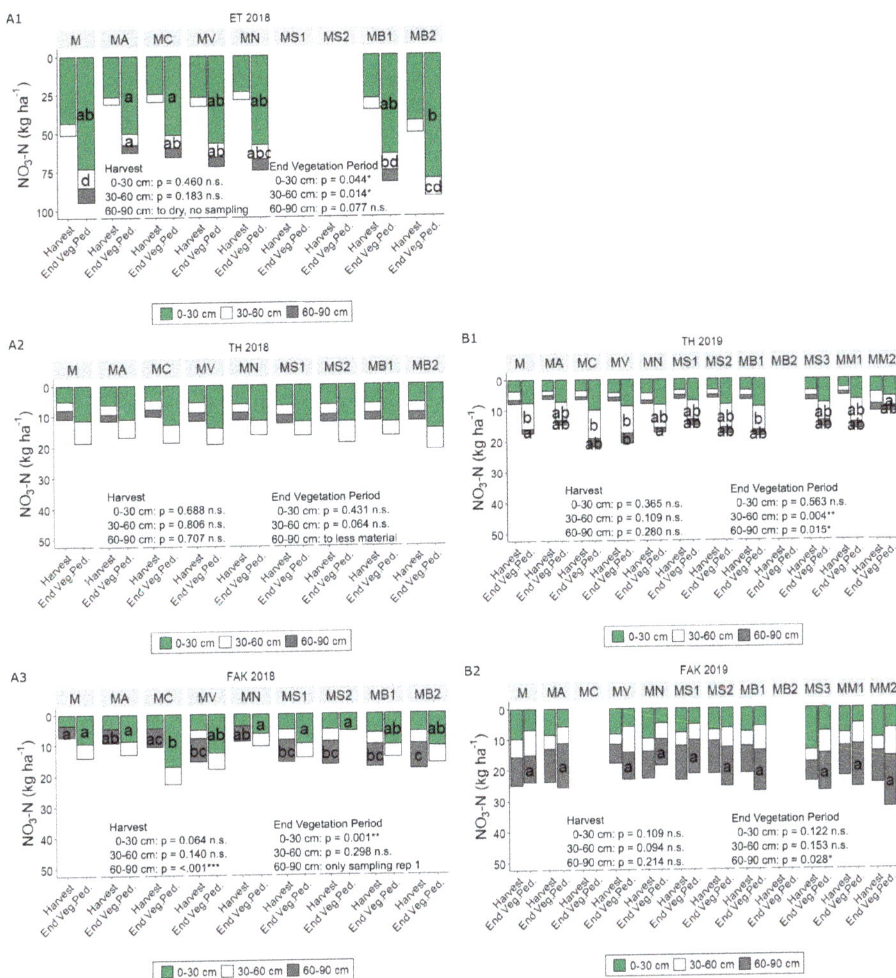

Figure 1. NO$_3$-N (kg ha^{-1}) of soil samplings for the different intercropped flowering partner (IFP) treatments (M Maize (Control), MA Maize + Alfalfa, MC Maize + Yellow sweet clover, MV Maize + Vetch, MN Maize + Nasturtium, MS1 Maize + Summer Squash I, MS2 Maize + Summer Squash II, MB1 Maize + Common Bean I, MB2 Maize + Common Bean II, MS3 Maize + Summer Squash III, MM1 Maize + Mixture I, MM2 Maize + Mixture II) in Ettlingen (ET) (2018 **A1**), Tachenhausen (TH) (2018 **A2**, 2019 **B1**) and Forchheim am Kaiserstuhl (FAK) (2018 **A3**, 2019 **B2**) after harvest and at the end of the vegetation period (End.Veg.Ped.; =end of growing season, when the weather causes a plant growth stop) for the three depths 0–30 cm (green bars), 30–60 cm (white bars), and 60–90 cm (gray bars). Bars with the same letter within one depth and one sampling date indicate non-significant differences (HSD-test, α = 5%).

The N-level had a significant influence on the content of CP in the biomass at harvest time ($p < 0.001$) (Appendix A Table A1). With increasing N-level there was a significant increase in CP. The highest content was achieved at 100% N-level, with 8.91%. The placement of the IFP seeds did not have an influence on CP ($p = 0.236$), while the influence of IFP was significant ($p < 0.001$). The only statistical influence between the IFP could be found between common vetch (MV), with 9.97% and all the other IFP (Figure 2). Control (M) achieved 7.08%. In 2018 at TH, nasturtium (MN) and both summer squashes (MS1, MS2) had lower CP contents than the control. In the second season, nasturtium (MN) and summer squash I and III (MS1, MS3) showed once again significantly reduced CP contents compared to the control (6.09%), while mixture II (MM2) showed an increase.

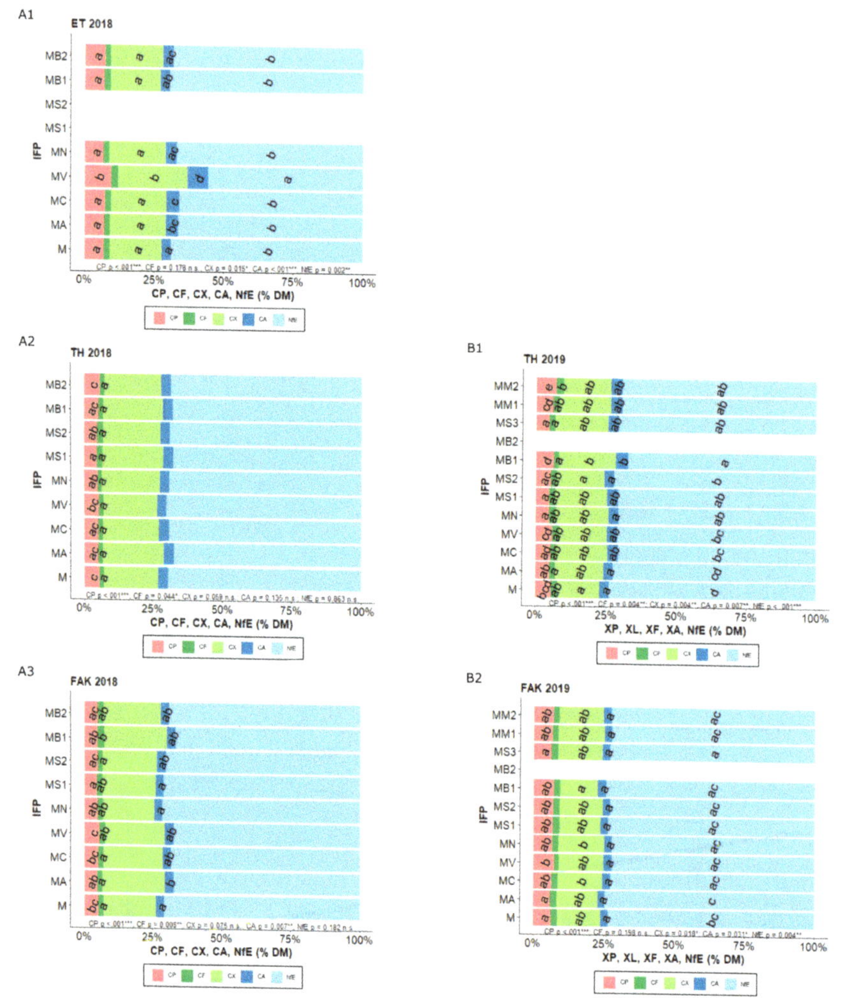

Figure 2. CP (crude protein), CF (crude fat), CX (crude fiber), CA (crude ash), and NfE (nitrogen-free extracts) content (% of DM) for the different intercropped flowering partner (IFP) treatments (M Maize (Control), MA Maize + Alfalfa, MC Maize + Yellow sweet clover, MV Maize + Vetch, MN Maize + Nasturtium, MS1 Maize + Summer Squash I, MS2 Maize + Summer Squash II, MB1 Maize + Common Bean I, MB2 Maize + Common Bean II, MS3 Maize + Summer Squash III, MM1 Maize + Mixture I, MM2 Maize + Mixture II) in Ettlingen (ET) (2018 **A1**), Tachenhausen (TH) (2018 **A2**, 2019 **B1**), and Forchheim am Kaiserstuhl (FAK) (2018 **A3**, 2019 **B2**). Bars with the same letter within one parameter indicate non-significant differences (HSD-test, $\alpha = 5\%$).

For FAK in both years, CP differences were measured (Figure 2), while in 2018, the summer squash I (MS1) had lower CP contents than the control. In 2019, common vetch (MV) had higher CP contents. No other intercropping treatment showed significant changes.

Neither the N-level ($p = 0.407$, $p = 0.694$, and $p = 0.324$) nor the seed placement ($p = 0.854$, $p = 0.725$, and $p = 0.870$) had significant influences on CF, CX, and CA content in ET 2018 (Table A1). The IFP also did not have an influence on the CF content ($p = 0.178$) (Figure 2). Common vetch intercropping (MV) showed higher CX contents than the control (M), while nasturtium (MN) and both beans (MB1, MB2) had comparable CA contents with the control. The NfE content was influenced by the N-level ($p = 0.003$) but not by the seed placement of the IFP ($p = 0.959$) (Appendix A Table A1). A high fertilization leads to lower NfE contents. The NfE behaved in the opposite way to the CP. Common vetch (MV) showed the significant lowest NfE contents in ET (Figure 2). In TH there were no significant differences between the IFP in 2018 for CF, CX, CA, and NfE, while in 2019, significant differences were observed for all parameters. For CF, none of the IFP differs from the control, while for CX and CA common bean I (MB1) had significantly higher amounts. For the NfE, all IFPs, except alfalfa intercropping (MA), had lower contents than the control. In FAK, only CF ($p = 0.008$) and CA ($p = 0.007$) showed significant differences in 2018. For CF, common bean I (MB1) showed higher contents than the control (M), while for CA, alfalfa (MA) had higher contents. In the second year, CX, CA, and NfE ($p = 0.018$, $p = 0.031$, and $p = 0.004$) showed significant differences, but none of the IFP was different from the control, except the summer squash III (MS3), which had a lower amount of NfE.

For use as feedstock in biogas plants, besides the DMY, the amount of produced biogas and methane are important factors. Only the N-level had a significant influence on the biogas yield ($p < 0.001$) (Appendix A Table A1). Significant decreases in biogas yields were found for 100% N fertilization compared to 0 and 50%, respectively, while the N-Level had no influence on methane yield ($p = 0.714$). Placement of IFP seed showed no influence on biogas and methane yield ($p = 0.546$ and $p = 0.733$) (Appendix A Table A1). The IFP had an influence on these two parameters in ET 2018 ($p = 0.025$ and $p = 0.016$) (Figure 3). Common vetch (MV) showed significantly lower biogas and methane yields, compared to the control (M). While the control achieved yields of 556 L kg^{-1} oDMC (organic DMC) for biogas and 300 L kg^{-1} oDMC for methane, intercropping with common vetch (MV) reduced these values to 539 L kg^{-1} and 294 L kg^{-1} oDMC, respectively. For TH, the response of the IFP on biogas and methane yield depended on the experimental year, while in 2018, biogas and methane yield were significantly increased by intercropping with common bean I (MB1), and in 2019, both yields were significantly decreased by intercropping with common bean I (MS1). In FAK, there were no significant differences on the biogas or methane yield for any of the IFP used, neither in 2018 nor in 2019.

In dairy cattle feeding, GE, ME, and NEL are important factors for evaluating the quality of the silage. At ET in 2018, there was only a significant influence of the N-level on the content of GE ($p = 0.021$), ME and NEL were not influenced by the level of N fertilization. The seed placement of the IFP had no influence on the feeding quality parameters (Appendix A Table A1). All three parameters showed that intercropping with common vetch (MV) significantly decreased the contents in ET.

In TH in 2018, ME and NEL showed significant differences, but no differences were found between the control (M) and any of the other IFP. The only significant difference was found between summer squash I intercropping (MS1) and common vetch intercropping (MV) (Figure 4). In 2019, except for the alfalfa intercropping (MA), all intercropping treatments had significant lower GE contents than the control (M). For ME and NEL, only intercropping with common bean I (MB1) showed significantly decreased contents. In FAK, no differences were observed in 2018 for any parameters. In 2019, only the GE showed significant changes. While intercropping with alfalfa (MA), yellow sweet clover (MC), common vetch (MV), and nasturtium (MN) did not differ significantly from the control (M), the use of the other IFPs significantly reduced the GE.

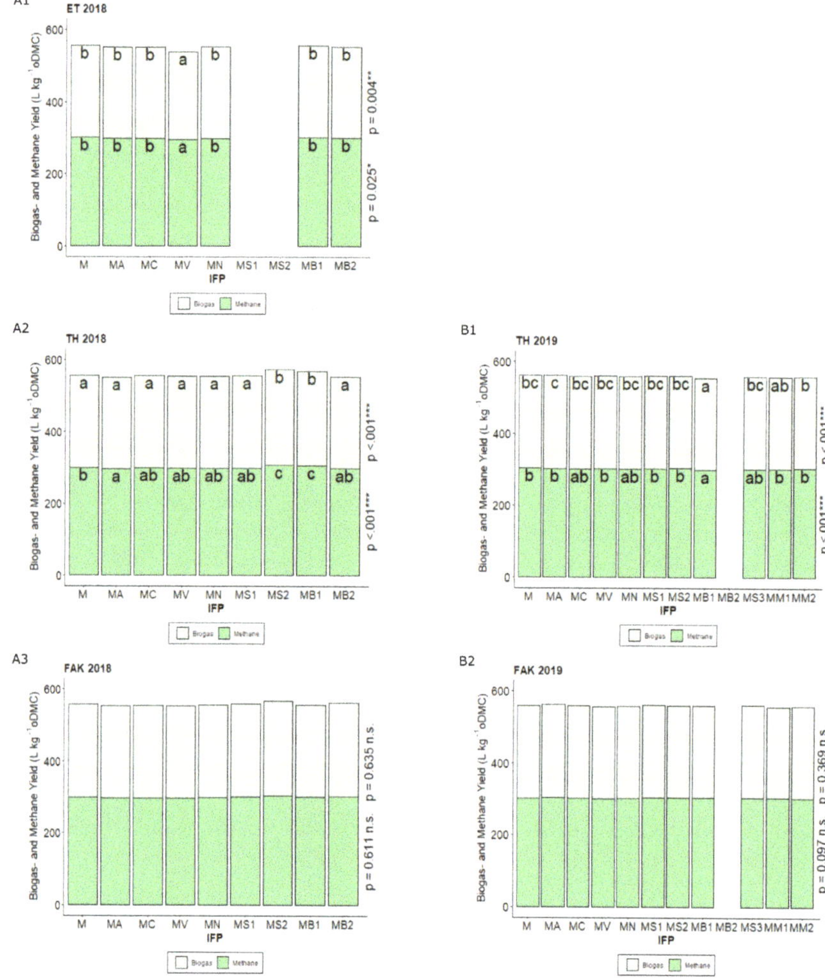

Figure 3. Biogas and Methane yields (L kg^{-1} oDMC) for the different intercropped flowering partner (IFP) treatments (M Maize (Control), MA Maize + Alfalfa, MC Maize + Yellow sweet clover, MV Maize + Vetch, MN Maize + Nasturtium, MS1 Maize + Summer Squash I, MS2 Maize + Summer Squash II, MB1 Maize + Common Bean I, MB2 Maize + Common Bean II, MS3 Maize + Summer Squash III, MM1 Maize + Mixture I, MM2 Maize + Mixture II) in Ettlingen (ET) (2018 **A1**), Tachenhausen (TH) (2018 **A2**, 2019 **B1**), and Forchheim am Kaiserstuhl (FAK) (2018 **A3**, 2019 **B2**). The upper end of each colored section in the stacked bars shows the respective value for the parameter's biogas and methane. Bars with the same letter within one parameter indicate non-significant differences (HSD-test, $\alpha = 5\%$).

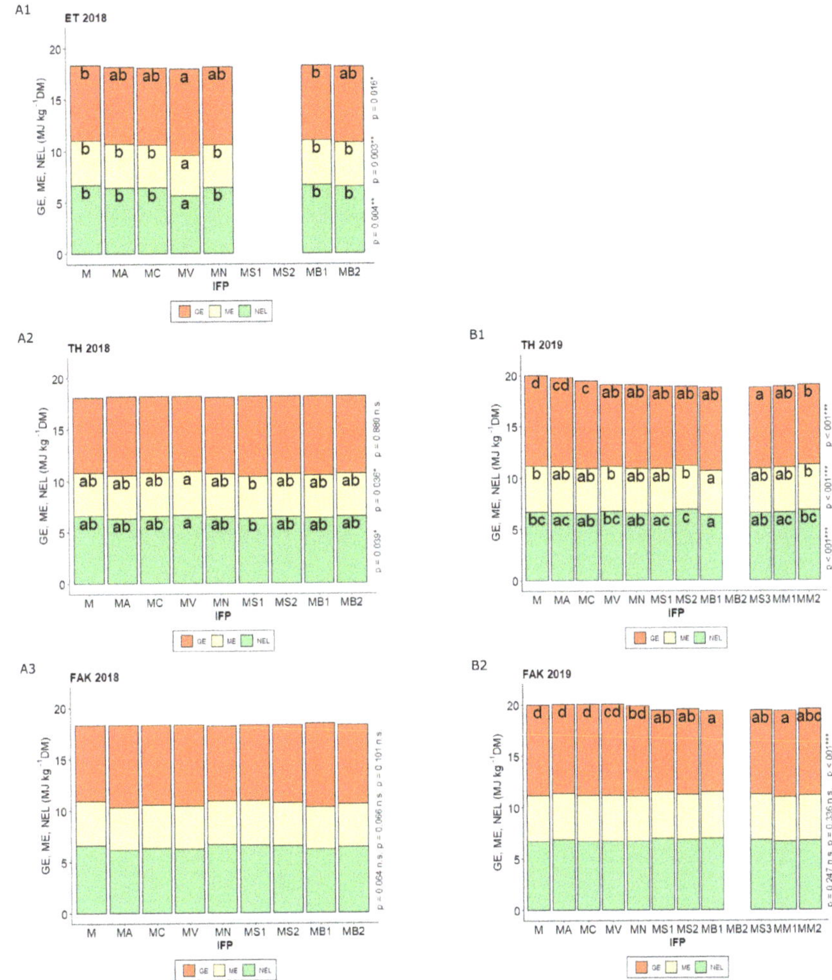

Figure 4. GE (gross energy), ME (metabolizable energy), and NEL (net energy for lactation) (MJ kg^{-1} DM) for the different intercropped flowering partner (IFP) treatments (M Maize (Control), MA Maize + Alfalfa, MC Maize + Yellow sweet clover, MV Maize + Vetch, MN Maize + Nasturtium, MS1 Maize + Summer Squash I, MS2 Maize + Summer Squash II, MB1 Maize + Common Bean I, MB2 Maize + Common Bean II, MS3 Maize + Summer Squash III, MM1 Maize + Mixture I, MM2 Maize + Mixture II) in Ettlingen (ET) (2018 **A1**), Tachenhausen (TH) (2018 **A2**, 2019 **B1**) and Forchheim am Kaiserstuhl (FAK) (2018 **A3**, 2019 **B2**). The upper end of each colored section in the stacked bars shows the respective value for the parameter's GE, ME, and NEL. Bars with the same letter within one parameter indicate non-significant differences (HSD-test, $\alpha = 5\%$).

4. Discussion

4.1. Influence of IFP on Maize Cropping

The following discussion will clarify the suitability of each IFP and offer suggestions for maize intercropping.

The single IFP provided a flower abundance for different periods of time and offer a food source for different insect species. This can make a major contribution for biodiversity promotion. With a flowering start before maize anthesis, alfalfa is an interesting partner to increase the flower abundance in maize crops. The literature showed that alfalfa is pollinated by wild bees, provides a pollen source for 29 species of wild bees and is a food resource for bumblebees [45–47]. Therefore, maize–alfalfa intercropping creates a flower/food abundance over a very long period. Yellow sweet clover per se is a pollen source for six wild bee species [46], which would have made this IFP interesting for biodiversity conservation. However, the results showed that yellow sweet clover did not flower at any location during the two years of the experiment. Literature showed that yellow sweet clover is not able to flower when shaded [48]. Additionally, this species has single- and two-year genotypes; the single year genotypes will already flower in the year of sowing and the two-year genotypes will only flower in the second year [49]. Since sweet clover is a plant species that has hardly been researched in breeding, no guarantee can be provided which flowering genotype or which mixing ratio the seed contains. The earliest flower abundance after sowing was supplied by common vetch, even earlier than alfalfa. The flowering period ends at about the same time as maize anthesis starts. Common vetch is pollinated by wild and honeybees, bumblebees, and is also a food resource for bumblebees [45–47]. Nasturtium, with a long flowering period which continued after the silage maize harvest, provides an interesting aspect for biodiversity. It will flower until the first frost or until soil tillage [50]. Therefore, it might be a habitat over a long time. A maize–nasturtium crop flowers 50 DAS, two weeks before the flowering of a sole maize crop. Nasturtium is pollinated by honey and wild bees, and also *Syrphidae* used it as a host [51]. In the region of maize origin, the traditional cropping system for maize was a combined cultivation of maize with summer squashes, common beans and others, the so called MILPA system. The summer squashes are pollinated by honey and wild bees and tolerate some shading [45,52], which makes them interesting for maize intercropping for biodiversity reasons. Flowering starts quite late, 1–2 weeks before maize anthesis, but as already observed for nasturtium, it continues even after harvest, until frost or soil tillage. Intercropping of maize and common beans is already done in practice. In 2019, 4000 ha were cultivated in Germany [53]. Common beans as the other main partner in the MILPA system are mainly self-pollinating, although insect pollination can enhance the seed yield [54–56]. Wild bees, butterflies, flies and bugs were observed to pollinate common beans [57]. The two-year experiments showed that common beans tend to a quite late flowering start. Mostly they started flowering after the maize started. Both mixtures (MM1, MM2) combined the mentioned flower abundance characteristics of two different IFP; summer squash II and common bean I (MM1), common vetch and common bean I (MM2). Especially for MM1 this leads to a prolonged flower abundance after end of maize anthesis by the squash, MM2 leads to an early flower abundance due to the common vetch.

An important aspect in maize intercropping is a system adapted weed control. Weed compete with maize (a water requiring crop) for nutrients and water, especially in dry years. In most tropic and sub-tropic countries, where intercropping in smallholder farm systems is the main agricultural practice, weed management is done by hand. However, if maize intercropping should take place on larger scales for a high economic performance, a weed control by common agricultural techniques must be possible. If weed control cannot be done, weeds form a major competition for maize, especially at sites with a high weed infestation. This has been shown for several IFP. Intercropping maize with alfalfa or sweet yellow clover leads to a weed share of 12.5% and 10.8% at ET, mainly consisting of *C. album*. ET is a typical maize location. Redwitz and Gerowitt [58] showed that fields that had, in previous years, a high share of maize in the crop rotation have an increased potential for infestation with *C. album*. The chemical weed treatment under conventional conditions consisted of a reduced amount of *Pendimethalin*. According to the manufacturer, the practical application rate of *Pendimethalin* should not be less than 3.5 L ha^{-1} (455 g a.i. L^{-1}) for an adequate *C. album* control in both pre- and post-emergence. It is also highly recommended that the application should take place not later than the three-leaf stage of the broadleaved weeds; most effective is the pre-emergence application [59]. Additionally, at TH the weed potential was not that high as in ET. Under organic conditions hoeing covered the alfalfa and the sweet yellow clover with soil, which the small plants did not tolerate.

Very few or none of the plants managed to grow from the heaped-up soil. The opposite was observed for maize–common vetch intercropping, where an adequate weed management is only possible under organic management. The hoeing in FAK worked well and the common vetch showed no difficulties growing out of the heaped-up soil, otherwise than alfalfa or yellow sweet clover. Under conventional management, especially at ET with its high weed infestation potential, weed management was difficult. In the first experimental year at ET, maize–common vetch plots had a high infestation with *C. album*. Although the same herbicide application as for maize–alfalfa and maize–yellow sweet clover took place in the second year in post-emergence, weeds could not be controlled successfully, for the reasons stated above. At TH, with a low infestation potential, the weed control used worked well. The advantage of nasturtium over alfalfa, yellow sweet clover and common vetch is the availability of an adapted weed management by a pre-emergence application of *Pendimethalin*, which controlled the weed infestation better, than a post-emergence application. The reduced amount of active ingredient can be a disadvantage in this intercropping system, especially on sites with a high weed potential. Under organic weed management, the nasturtium showed no negative effect when buried under heaped-up soil during mechanical weed management. Afterwards, the nasturtium plants grew up from the heaped-up soil. Plant protection in maize–squash intercropping is practicable under organic conditions, while the chemical plant protection is challenging. Most of the registered maize herbicides, which are also allowed in squashes, can only be applied as inter-row application, which requires special equipment. For common beans, mechanical weed control or the application with *Pendimethalin* and *Dimethenamid*-P enabled an effective weed control.

To achieve high DMY, a good maize development and growth must be enabled. An important growth parameter for high DMY is the final plant height of the maize [60–64]. Intercropping of maize with alfalfa mostly showed no differences in plant height, except for TH in 2019. Studies from Canada also showed no change in plant height by intercropping maize with alfalfa compared to a sole maize stand [60]. This was confirmed by the DMY results 2019 in TH. On the other hand, ET showed a reduction in DMY but no change in plant height. This can be attributed to the high weed infestation (due to the reduced amount of active ingredient and late application date) and the resulting competition by a high weed share [65]. For grain maize–yellow sweet clover intercropping, Abdin et al. [60] showed that no reduction in plant height is expected. This was confirmed by our results, only TH 2019 forms an outlier with a significantly reduced maize plant height. In maize–yellow sweet clover, plant height was not affected by IFP, but the same effect on weeds was observed as mentioned above under maize–alfalfa intercropping. The non-significant change in plant height and DMY also make nasturtium an interesting partner for intercropping in maize. Also, no changes in DMY compared to sole maize cropping were found for intercropping maize with common beans. Experiments from Northern Germany also showed no differences in DMY [35]. In Great Britain, no significant differences in DMY between maize (10 plants m^{-2}) and maize–bean (7.5:5 plants m^{-2}) were observed [31]. In contradiction to the four previously mentioned intercropping treatments, intercropping with common vetch showed reduction in the final plant height of maize, depending on the year. The reduction in plant height was not significant in the first year at the locations in our study. In the second-year, significant height reductions for maize plants intercropped with common vetch were observed at TH and FAK. Also, the DMY was reduced. Common vetch is a plant which leaches allelopathic substances during decomposition of its biomass [66]. In 2018 it was observed at all locations that, after a short time period, the vetch was infested with mildew and died. Root excretions and leaching's (vanillin acid, p-coumaric acid, ferulic acid) of above-ground common vetch biomass is shown to inhibit the germination of wheat, as well as plant growth and development [66]. Their study concluded that wheat produced more root biomass than above ground biomass, due to the allelopathic substances. This change in above/below biomass ratio is a reaction which should promote plant growth. Since these effects has also been observed in rice [67], it is obvious that such effects also might occur in maize, which, like wheat and rice, belongs to the Poaceae. The squashes showed different growth behavior and influence on DMY, which indicates a great variability in the species, which could be shown by squash III in FAK 2019. Squash I and II showed no change in maize plant height and DMY, but squash III in FAK reduced the final maize plant height. While squash I and II produced fruit weights of 0.1 and 0.2–0.4 kg respectively, squash III can achieve fruits of 2–4 kg [68]. The production of fruits with a high weight requires nutrients and water, which are no longer available for maize. To the best of our knowledge, there are no experiments on maize–squash intercropping for silage usage. Most maize–squash experiments are done with maize for grain production. A study from traditional smallholder farm systems in Africa showed, that there were no differences in grain yield by intercropping with squashes; as long as the sowing rate of squashes meets 20% of the maize sowing rate [69]. Our experiment used 1.6 seeds m^{-2}, which equals 20% of the sowing rate of 8 maize seeds m^{-2}. Therefore, no changes were expected, which was proved by no changes in DMY. But the study from Mashingaidze et al. [69] also showed, that no differences in grain yields can only be achieved in years where a high infestation with *Sphaerotheca fuligniea* took place. The infestation inhibits the formation of fruits. In years without infestation, fruits are formed, and the squash plants compete with maize for nutrients. This leads to the conclusion that larger fruits have a higher competitiveness and, therefore, a greater influence on maize. This is in line with the results of squash III. This squash formed bigger fruits than the other two squashes and showed significant reductions in final DMY of the MSIII plots and yield relevant parameters. Also, the sowing of a mixture of maize and squash is a challenge. The 1000 grain-mass varies widely within the genus Cucurbita. While squash I and II had a 1000-grain mass of 55 g and 27 g, respectively, squash III had 102 g and was suitable for a combined seeding with maize as a mixture.

Prevention of the water bodies due to soil NO_3-N leaching by IFP could not be fully proven. Under maize–alfalfa intercropping the soil NO_3-N content was significantly reduced in the layer of 30–60 cm depth compared to the control at vegetation end in ET 2018. It could be observed that the alfalfa continued growing after harvest. It is therefore reasonable to assume that the IFP continued absorbing soil NO_3-N. Although, nasturtium continues to grow after harvest and provides a flower abundance, it did not show significant differences in the soil NO_3-N content in the single experimental years or sites, except for a reduction in 30–60 cm at the end of the vegetation period in ET 2018.

The IFP or a 'pollution' by a high weed share can also cause changes in the chemical composition of the harvested material. Most often the CP and CA content were changed. For biogas production and cattle feeding, the following important parameters should also not change to a disadvantageous content: CP (too long decomposition time in biogas plant, in dairy cattle to much CP causes high milk urea concentration, or CP is sometimes degraded to fast and, therefore, has a low utilization), CX (difficult digestion in biogas plant and dairy cattle stomach), CA (reduction in digestion space of the biogas plant and no substances that can be converted for energy and feeding purpose). An increase in NfE has no negative effects, due to the fact that highest methane concentration can be achieved by this parameter and cows can convert most energy from these, respectively. The matrix of correlation coefficients in ET for 2018 showed that there is a negative correlation of CP on the NfE content ($r_p = -0.68$) (Figure 5). This trend can be explained due to the fact, that NfE are the nitrogen-free extracts. If there is an increase in CP, less amount of the dry matter can consist of NfE. This was proven 2018 in ET where intercropping with common vetch increased the CP contents significantly, while the NfE decreased significantly. Such a trend could also be observed in TH 2019 for mixture I (MM2), which also contains vetch. CP can build high contents of methane, but it is a slowly degradable ingredient in biogas plants. This has been shown by the correlation coefficients for biogas ($r_p = -0.76$) and methane ($r_p = -0.47$). A high content in CP reduces the biogas contents due to the long retention time. But the higher the content of CP, the more methane could be built. Also, the CA content influences the final yields of biogas and methane. If the content of CA in the biogas substrate is too high, ash can settle at the bottom of the fermenter, reducing the digestion space and thus the yields. Therefore, the matrix showed a significantly negative influence of CA on biogas ($r_p = -0.90$) and methane ($r_p = -0.75$). Results from 2019 in TH verified these findings. An increase in CA by intercropping with common bean I leads to increased CA contents and decreased biogas and methane contents. The nutrition parameters GE,

ME and NEL are mainly influenced by CP and CX. CX showed a negative influence on GE ($r_p = -0.50$), ME ($r_p = -0.99$) and NEL ($r_p = -0.99$), CP had a negative influence on ME ($r_p = -0.50$) and NEL ($r_p = -0.50$). This negative influence of CX was also proved, in 2019 in TH, an increase in CX by intercropping with common bean I significantly reduced GE, ME, and NEL due to the low nutritive value of CX.

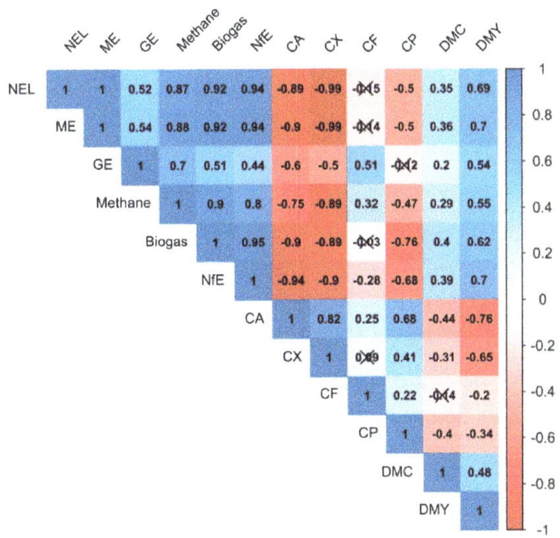

Figure 5. Pearson's coefficient of correlation within and across yield and quality parameters at ET in 2018, averaged over the parameters N-Level, seed placement of the intercropped flowering partners (IFP) and IFP. Non-significant coefficients ($p < 0.05$) are crossed out.

Significantly higher CA contents were observed for ET and FAK in 2018 under maize–alfalfa, maize yellow sweet clover and maize–common vetch intercropping. This can be attributed to the high share of weed in the harvested biomass. Most of the weed consisted of *C. album*, which has a CA content of 23.3% [70]. Adedapo et al. [70] also showed that *C. album* has high contents of CX (16.7%) and CA (23.3%). An increase in these two parameters has negative effects on biogas and methane yield, and on the feed quality parameters, GE, ME and NEL. This was proven by the significant reduction in biogas, methane, GE, ME and NEL in ET 2018. High contents of CA can disturb the biogas process by settling down on the bottom of the fermenter and reduce the space for digestion. In addition, CA does not provide a basis that can be converted into energy, neither in biogas plants nor in cattle feeding. These effects were proven by the highly negative correlations between CA and CX on biogas, methane, GE, ME and NEL shown in Figure 5. Biogas and methane formation depend on the composition and biodegradability of the substrates used [71]. CX did not contribute much to methane yield [72].

The increased CP content could also be attributed to the high weed infestation. In ET 2018, the weed covered area was higher than 50%. *C. album* has a high CP content [73,74]. Depending on plant age, *C. album* can accumulate an additional 1.75–5.27% nitrogen, which corresponds to 10.9–32.9% CP [73,74]. The *C. album* will be chopped together with the maize. A study by Sarabi et al. [75] showed that the control of *C. album* in maize is absolutely necessary in order to prevent growth inhibition and prevent yield losses. Especially for common vetch intercropping, root extracts from the vetch can increase the CP content. Aarssen et al. [76] showed in an intercropping trial with *Avena sativa* and common vetch that the vetch root extracts increased the nitrogen content of *Avena sativa*. These changes were also confirmed by the 2019 results of TH and FAK. In FAK, an increase in CP content was measured, which resulted in a reduced NEL. In contrast, a decrease of the NfE content was measured in TH, which corresponds to an increase in CP and also achieved reduced NEL contents.

The above-mentioned studies from northern Germany and Great Britain showed increased CP contents at unchanged DMY. However, our study could not prove an increase in CP content by intercropping with common beans. This could be ascribed to the sown maize:bean ratio and the weather conditions. The studies from northern Germany and Great Britain used a higher bean and a lower maize proportion. While in our study the sown maize:bean ratio of 8:4.5 plants m^{-2} resulted in a percentage share of 64% maize and 36% beans seeds, Fischer et al. [35] used maize:bean ratios of 8:6, resulted in a share of 57% maize and 43% beans and Dawo et al. [31] a 7.5:5 ratio (60% maize and 40% bean) and a 5:5 ratio (50% maize and 50% bean). Most studies which proved a higher CP content sown a higher proportion of beans. An additional reason for the missing CP increase in our study, could be the environmental conditions in Southern Germany. Common beans are sensitive to high temperatures. Growing at 32 °C decreases the biomass of *P. vulgaris*, compared to growing at 25 °C [77,78]. At full flowering, temperatures above optimum (15–30 °C) were shown to affect all enzyme activities of the nitrogen metabolism negatively [79]. This leads to the conclusion that the temperatures during flowering in southern Germany can be above the optimum, which influences the nitrogen metabolism and therefore the formation of CP, while in the cooler climates of northern Germany and Great Britain the nitrogen metabolism is less affected. This was also shown by Porch and Jahn [80]. They showed that at high day/night temperatures (32/27 °C) heat-sensitive *P. vulgaris* genotypes react with an excessive abscission of their reproductive organs. This gives an explanation, why in hot, dry 2018 no effect on CP could be observed. Another reason for the lack in increase in CP, even in 2019, may be the proportion of beans in the silage. This should be above 20% [81]. This increase could not be achieved at any site in any year (Tables 4–6). In addition, it must be noted that different varieties of *P. vulgaris* or species of Phaseolus beans were used in the above-mentioned experiments. Depending on the genotype used, beans showed a wide range in CP contents. Celmeli et al. [82] showed that the CP content in both landraces and modern cultivars of common beans varied widely (landraces: 16.5–25.2%, modern cultivars: 19.7–24.3%). Thus, in addition to the seeding rate and growth conditions, the selected bean variety can have a significant influence on the CP content of the harvested material. In 2019, the results from TH showed a decrease in biogas and methane, while CX and CA increased and NfE decreased. The correlation matrix in Figure 5 verifies this finding. High CA contents had a negative correlation on biogas and methane (r_p = −0.90 and −0.75), also high CX (r_p = −0.89 and −0.89). CA and CX are not or only slowly digestible in the biogas plant, while NfE had a positive correlation on biogas and methane (r_p = 0.95 and 0.80), respectively. Therefore, the decreasing NfE content leads to decreasing biogas and methane yields. The reduced contents in GE, ME and NEL also could be explained by these increases in CA and CX. Mixture I showed neither in TH nor in FAK differences in DMY or CP. Mt. Pleasant [83] also showed that traditional MILPA systems did not reach higher yields or an increase in CP compared to a sole maize crop. Mixture II (common vetch and common bean II) had significantly lower DMY but higher CP and lower NfE compared to the control. These changes can be ascribed to the mentioned negative effects of common vetch intercropping. Decomposing vetch parts inhibit maize growth (=reduced plant height), while the combination of two legume IFP can increase the CP content. Also, the high weed infestation increased the CP content.

4.2. Requirements on Agricultural Practice and Equipment

Intercropping of maize and IFP has several challenges, like seeding technique and weed control, which will be discussed in detail in the following contents.

When maize and an IFP are sown together in a single working step as mixture via a single seed precision planter, the morphology of the seeds should be similar, e.g., shape, 1000-grain mass. This enables sowing as a seed mixture. A single-step sowing reduces the demand of fossil fuels, reduces soil compaction and saves machinery and labor costs. However, sowing as a mixture requires that both maize and IFP can deal with the weed management. Adapted weed control is important, otherwise there will be too much competition which will lead to yield losses. As seen by Abdin et al. [44], cover crops were not able to suppress weed development adequately, especially on sites with high weed potential. This has been shown in ET in 2018. Reduced amounts of the recommended plant protection agents for maize were used in order to not harm the IFP. These reduced amounts lead to a high pressure of problematic weeds, such as *C. album* (L.) and *Galinsoga* spp. (RUIZ & PAV.). Small IFP seeds, like alfalfa, yellow sweet clover, common vetch, and squash I and II are technically not suitable for sowing as a mixture. These IFP also did not permit an adequate chemical weed control under the local site conditions. Some IFP's had under mechanical weed control problems by burying during hoeing. The only practicable method could be a sowing of IFP via box spreader during the last mechanical weeding step.

Separate sowing enables on the one hand the establishment of a spatially separated crop stand and on the other hand, it is also possible to establish the plants at different times. Thus, maize is sown first and weed control can be done (chemically or mechanically). Afterward, IFP are sown in the maize crop. However, this delays the flowering of the IFP. In case of a spatially separated establishment of the IFP between the maize rows, no mechanical weed control can take place, if the IFP seed is sown immediately after the maize sowing. In both cases, establishment in a second step is necessary, which costs time and fossil resources. Special machines for maize and IFP sowing in one step are not yet widely used. A box spreader on the maize seed drill also means that no mechanical inter-row weed treatment can take place. It should be clear, when sowing as mixture, the use of a partner additionally reduces the share of maize in the crop, if the proportion of maize in the seeding mixture is not increased. This is caused by the seed separation at the sowing disc of the precision planter. The seeds will randomly get into the holes of the seed disc. This means that at the adjusted sowing rate (e.g., 10 plants m^{-2}), it is not possible to identify exactly the distribution pattern of maize and IFP within the row, this will happen randomly. Since maize forms the major share of biomass for most of the observed IFP treatments, except for the common bean, it must be ensured by a high maize proportion and a low IFP proportion that the maize seed ratio is not "thinned" by the IFP. In general, if an intercropping mixture is sown in one working step, the proportion of the IFP should be kept as high as possible for biological promotion and as low as possible so the maize will not be "thinned".

5. Conclusions

Maize row-intercropping with flowering plants represents an alternative to sole maize cropping, due to the fact that cropping partners can mitigate negative environmental impacts. Some cropping partners proved to be unsuitable for row-intercropping in maize crops. Maize–alfalfa, maize–yellow sweet clover, and maize–common vetch cannot be recommended due to the difficulties in weed control and sowing as a seed mixture. Common vetch and mixtures which contain common vetch also are not recommended due to the assumed allelopathic effects of the decomposing biomass. Maize–squash could be an interesting alternative under mechanical weed control, while maize–nasturtium showed to be an alternative to sole maize cropping under conventional and organic farming, also the maize–bean crop could be an option. The yields and qualities of maize–nasturtium and maize–bean biomass are comparable with sole maize cropping, while the flowers provide a pollen source. Nasturtium could also be possible for grain maize production because no plant parts contaminate the grain. Intercropping maize with nasturtium requires additional studies on chemical weed control and the sowing rate of nasturtium. If a combined sowing of maize and IFP in a single step can be done, there will be no additional consumption of fossil fuels and no increased soil compaction compared to sole maize cropping. However, this requires that the seeds be sown as a mixture. This means that the seeds should be similar in their morphology. This can already be achieved by maize–bean. When maize and IFP are sown separately, plants can be used whose seed morphology differs from maize. In this case the seeds are also located between the maize rows and no mechanical weed control or inter-row band application of herbicides can take place. A later sowing of the IFP after finishing all weed control measures (chemical or mechanical), would be possible. However, this would delay the flowering time of the IFP, and an additional flower abundance before maize flowering can no longer be enabled. By sowing in one working step as a seed mixture, the share of IFP should be kept as low as possible to enable a high proportion of maize. The use of IFP must therefore be well planned a priori, in terms of the benefits for biodiversity, shade tolerance of the IFP, the competitive situation for maize (water, nutrients), the sowing technique, the possibility of weed control, and the possible influences on final biomass. It is suggested that maize intercropping, as already practiced by ambitious farmers, should be further explored by research.

Author Contributions: Conceptualization, K.S., C.S., M.M.-L. and K.M.; methodology, V.S.S., C.S., M.M.-L. and K.S.; software, V.S.S.; validation, V.S.S., M.M.-L. and C.S.; formal analysis, V.S.S.; investigation, V.S.S., C.S., M.M.-L. and K.S.; data curation, V.S.S. and C.S.; writing—original draft preparation, V.S.S.; writing—review and editing, V.S.S., M.M.-L., C.S., S.W., K.S. and K.M.; visualization, V.S.S.; supervision, V.S.S., M.M.-L. and C.S.; project administration, V.S.S., C.S., M.M.-L., K.S., S.W. and K.M.; funding acquisition, K.S. and K.M. All authors have read and agreed to the published version of the manuscript.

Funding: This research was funded by the Ministry of Rural Affairs and Consumer Protection Baden-Württemberg (MLR) within the project "Diversifizierung des Silo- und Energiemaisanbaus im konventionellen und ökologischen Landbau".

Acknowledgments: The authors would like to thank Jan E. Neuweiler of the State Plant Breeding Institute, University Hohenheim for statistical support; Conny Hüber for information about pollinators; Marwin Dauth, Khaliun Sukhbaatar, and Sabine Hubert for data collection and the staff of the LTZ Augustenberg for making this joint research project possible.

Conflicts of Interest: The authors declare no conflict of interest. The funders had no role in the design of the study; in the collection, analyses, or interpretation of data; in the writing of the manuscript, or in the decision to publish the results.

Appendix A

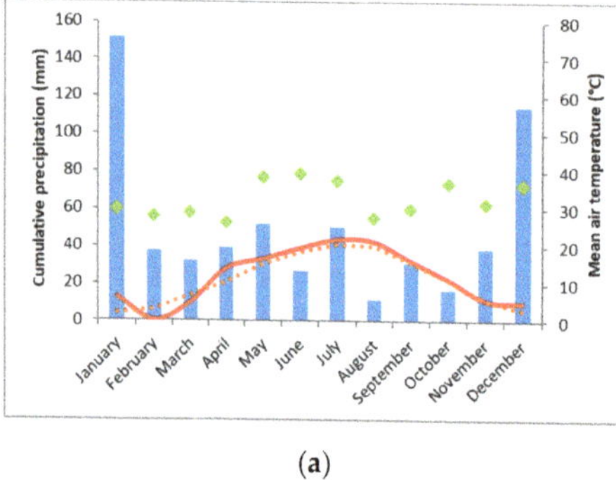
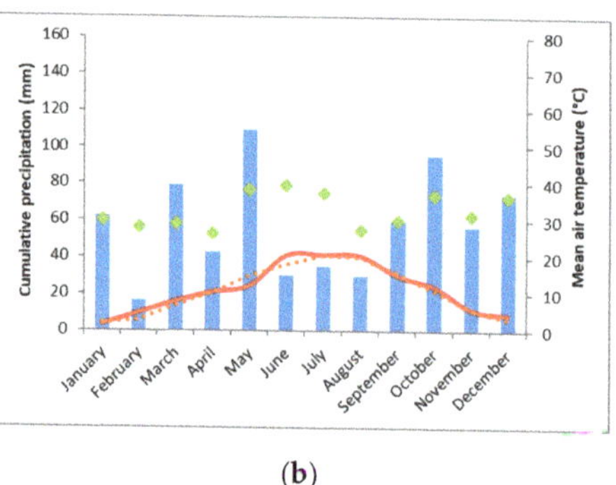

(a) (b)

Figure A1. Climate chart of the experimental location Ettlingen (ET) for 2018 (**a**) and 2019 (**b**). The monthly cumulative sum of precipitation (mm, blue bars), the average air temperature (°C, solid, red line), the mean long-term annual precipitation (mm, filled, green diamonds), and the average long-term air temperature (°C, dashed, orange line). (Data source: Meteorological station Rüppur, 3.5 km linear distance to experimental site).

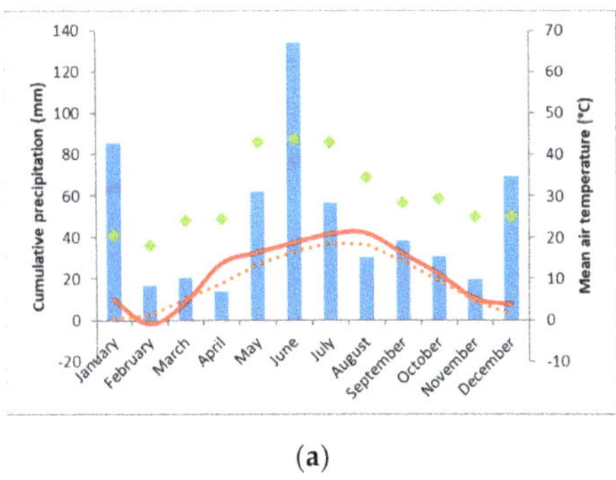 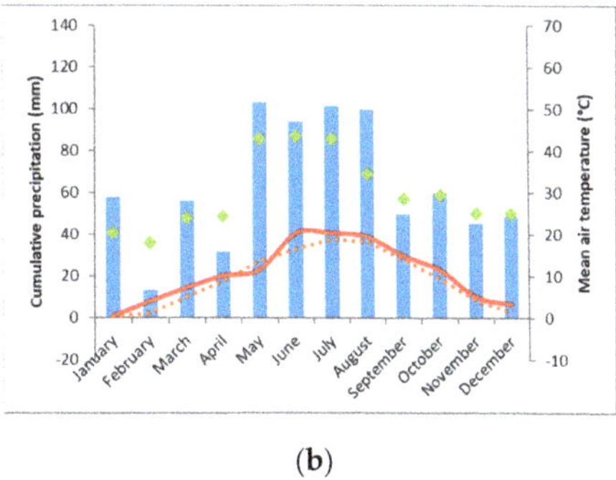

(**a**)   (**b**)

Figure A2. Climate chart of the experimental location Tachenhausen (TH) for 2018 (**a**) and 2019 (**b**). The monthly cumulative sum of precipitation (mm, blue bars), the average air temperature (°C, solid, red line), the mean long-term annual precipitation (mm, filled, green diamonds), and the average long-term air temperature (°C, dashed, orange line). (Data source: meteorological station Tachenhausen, 0.5 km linear distance to experimental site).

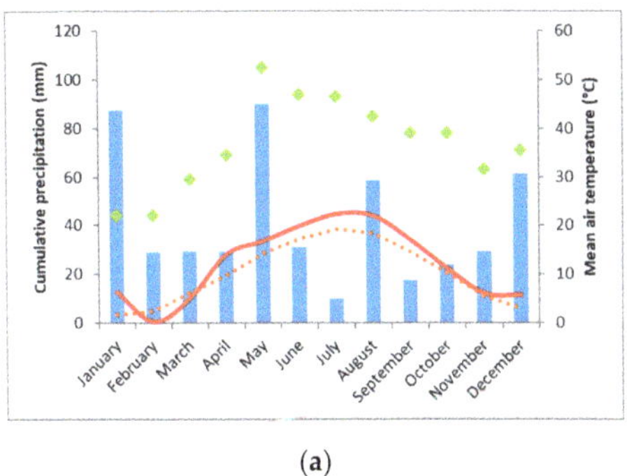 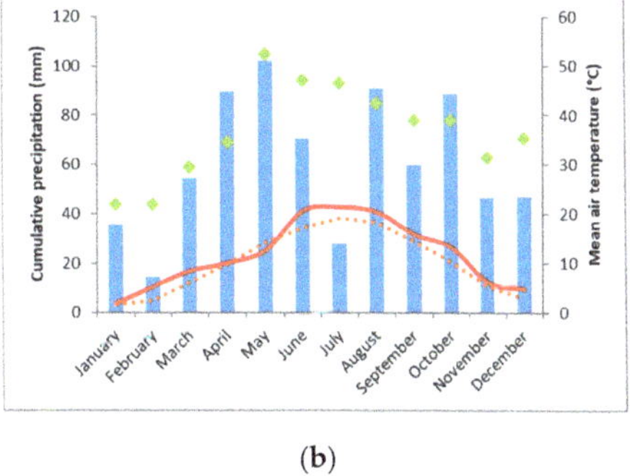

(**a**)   (**b**)

Figure A3. Climate chart of the experimental location Forchheim am Kaiserstuhl (FAK) for 2018 (**a**) and 2019 (**b**). The monthly cumulative sum of precipitation (mm, blue bars), the average air temperature (°C, solid, red line), the mean long-term annual precipitation (mm, filled, green diamonds), and the average long-term air temperature (°C, dashed, orange line). (Data source: meteorological station Herbolzheim, 9.3 km linear distance to experimental site).

Table A1. CP (crude protein), CL (crude fat), CX (crude fiber), CA (crude ash), NfE (nitrogen-free extracts) content (% of DM), biogas and methane yield (L kg^{-1} oDMC) and GE (gross energy), ME (metabolizable energy), and NEL (net energy for lactation) (MJ kg^{-1} DM) for the different N-Levels and seed placement of the intercropped flowering partners (IFP) in ET 2018.

	CP		CF	CX	CA	NfE		Biogas		Methane	GE		ME	NEL
				(% DM)				(L kg^{-1} oDMC)			(MJ kg^{-1} DM)			
N-Level														
0%	6.62	A	2.01	21.5	4.46	70.5	B	554	B	298	18.1	A	10.6	6.39
50%	7.83	B	2.23	21.6	4.65	68.9	AB	552	B	299	18.3	B	10.6	6.37
100%	8.91	C	2.09	21.2	4.73	68.3	A	549	A	298	18.3	AB	10.6	6.40
p-value	<0.001 ***		0.407	0.694	0.324	0.003 **		<0.001 ***		0.714	0.021 *		0.849	0.855
Seed placement														
BR	7.68		2.10	21.3	4.6	69.2		552		299	18.2		10.6	6.40
IR	7.89		2.11	21.6	4.63	69.5		551		298	18.2		10.6	6.37
p-value	0.236		0.854	0.725	0.87	0.959		0.546		0.733	0.842		0.754	0.763

Values with the same letter within one parameter indicate non-significant differences within the three factor levels (N-Level and Seed placement) (HSD-test, α = 5%). *** $p \leq 0.001$, ** $p \leq 0.01$, * $p \leq 0.05$.

45. McGregor, S.E. *Insect Pollination of Cultivated Crop Plants*; Agricultural Research Service, US Department of Agriculture: Washington, DC, USA, 1976; Volume 496.

46. Westrich, P. *Die Wildbienen Baden-Württembergs, 1: Allgemeiner Teil: Lebensräume, Verhalten, Ökologie und Schutz*; Ulmer: Ulm, Germany, 1989.

47. Schindler, M.; Schumacher, W. Auswirkungen des Anbaus vielfältiger Fruchtfolgen auf wirbellose Tiere in der Agrarlandschaft (Literaturstudie). In *Schriftenreihe des Lehr-und Forschungsschwerpunktes USL*; Universitäts- und Landesbibliothek Bonn: Bonn, Germany, 2007; Volume 147.

48. Ogle, D.; St John, L.; Tilley, D. *Plant Guide for Yellow Sweetclover (Melilotus officialis (L.) Lam.) and White Sweetclover (M. alba Medik.)*; USDA-Natural Resources Conservation Service, Idaho Plant-Materials Center: Aberdeen, UK, 2008.

49. Turkington, R.A.; Cavers, P.B.; Rempel, E. The Biology of Canadian Weeds: 29. *Melilotus alba* Desr. and *M. officinalis* (L.) Lam. *Can. J. Plant Sci.* **1978**, *58*, 523–537. [CrossRef]

50. Christenhusz, M.J. Tropaeolum Majus. *Curtis's Bot. Mag.* **2012**, *29*, 331–340. [CrossRef]

51. Comba, L.; Corbet, S.A.; Hunt, L.; Warren, B. Flowers, nectar and insect visits: Evaluating British plant species for pollinator-friendly gardens. *Ann. Bot.* **1999**, *83*, 369–383. [CrossRef]

52. Plants for A Future Cucurbita Pepo Pumpkin, Field pumpkin, Ozark Melon, Texas Gourd PFAF Plant Database. Available online: https://pfaf.org/user/plant.aspx?LatinName=Cucurbita+pepo (accessed on 29 October 2018).

53. Mund, F. Mais-Stangenbohnen—Ein Gemenge der Zukunft? Available online: https://www.topagrar.com/acker/aus-dem-heft/mais-stangenbohnen-ein-gemenge-der-zukunft-11860265.html (accessed on 3 November 2020).

54. Douka, C.; Nganhou, S.N.; Doummen, F.N.; Mengue, A.M.M.M.; Tamesse, J.L.; Fohouo, F.-N.T. Diversity of Flowering Insects and Their Impact on Yields of *Phaseolus Vulgaris* L. (Fabaceae) in Yaoundé (Cameroon). *J. Agric. Crop.* **2018**, *4*, 105–111.

55. Ramos, D.; Bustamante, M.M.C.; Silva, F.D.; Carvalheiro, L.G. Crop fertilization affects pollination service provision—Common bean as a case study. *PLoS ONE* **2018**, *13*. [CrossRef]

56. Raggi, L.; Caproni, L.; Carboni, A.; Negri, V. Genome-Wide Association Study Reveals Candidate Genes for Flowering Time Variation in Common Bean (*Phaseolus vulgaris* L.). *Front. Plant Sci.* **2019**, *10*. [CrossRef]

57. Kingha, B.M.T.; Fohouo, N.T.; Ngakou, A.; Bruuml, D. Foraging and pollination activities of *Xylocopa olivacea* (Hymenoptera, Apidae) on Phaseolus vulgaris (Fabaceae) flowers at Dang (Ngaoundere-Cameroon). *J. Agric. Ext. Rural Dev.* **2012**, *4*, 330–339.

58. Redwitz, C.V.; Gerowitt, B. Welche Faktoren fördern das Auftreten von *Chenopodium album* auf norddeutschen Maisflächen? *Jul. Kühn Arch.* **2014**, *443*, 165.

59. BASF. Stomp®Aqua. Available online: https://www.agrar.basf.de/de/Produkte/Produktdetails/Stomp%C2%AE-Aqua.html (accessed on 6 August 2020).

60. Abdin, O.; Coulman, B.E.; Cloutier, D.; Faris, M.A.; Zhou, X.; Smith, D.L. Yield and yield components of corn interseeded with cover crops. *Agron. J.* **1998**, *90*, 63–68. [CrossRef]

61. Gitelson, A.A.; Viña, A.; Arkebauer, T.J.; Rundquist, D.C.; Keydan, G.; Leavitt, B. Remote estimation of leaf area index and green leaf biomass in maize canopies. *Geophys. Res. Lett.* **2003**, *30*, 1248. [CrossRef]

62. Law-Ogbomo, K.E.; Remison, S.U. Growth and Yield of Maize as Influenced by Sowing Date and Poultry Manure Application. *Not. Bot. Horti Agrobot. Cluj Napoca* **2009**, *37*, 199–203.

63. Gao, S.; Niu, Z.; Huang, N.; Hou, X. Estimating the Leaf Area Index, height and biomass of maize using HJ-1 and RADARSAT-2. *Int. J. Appl. Earth Obs. Geoinf.* **2013**, *24*, 1–8. [CrossRef]

64. Li, W.; Niu, Z.; Chen, H.; Li, D.; Wu, M.; Zhao, W. Remote estimation of canopy height and aboveground biomass of maize using high-resolution stereo images from a low-cost unmanned aerial vehicle system. *Ecol. Indic.* **2016**, *67*, 637–648. [CrossRef]

65. Abouziena, H.F.; El-Metwally, I.M.; El-Desoki, E.R. Effect of Plant Spacing and Weed Control Treatments on Maize Yield and Associated Weeds in Sandy Soils. *Am. Eurasian J. Agric. Environ. Sci.* **2008**, *4*, 9–17.

66. Zohaib, A.; Tabassum, T.; Anjum, S.A.; Abbas, T.; Nazir, U. Allelopathic Effect of Some Associated Weeds of Wheat on Germinability and Biomass Production of Wheat Seedlings. *Planta Daninha* **2018**, *35*, 89. [CrossRef]

67. Zohaib, A.; Anjum, S.A.; Jabbar, A.; Tabassum, T.; Abbas, T.; Nazir, U. Allelopathic effect of leguminous weeds on rate, synchronization and time of germination, and biomass partitioning in rice. *Planta Daninha* **2017**, *35*, 32. [CrossRef]

68. KCB-Samen GmbH KCB Samen. Available online: https://www.kcb-samen.ch/ (accessed on 15 April 2020).

69. Mashingaidze, A.B.; van der Werf, W.; Lotz, L.A.P.; Kropff, M.J.; Nyakanda, C. Crop yield and weed growth in maize-pumpkin intercropping. In *Improving Weed Management and Crop Productivity in Maize Systems in Zimbabwe*; Tropical Resource Management Papers; Wageningen University: Wageningen, The Netherlands, 2004; ISBN 90-6754-820-0.

70. Adedapo, A.; Jimoh, F.; Afolayan, A. Comparison of the nutritive value and biological activities of the acetone, methanol and water extracts of the leaves of *Bidens pilosa* and *Chenopodium album*. *Acta Pol. Pharm.* **2011**, *68*, 83–92.

71. Bartuševics, J.; Gaile, Z. Influence of Maize Hybrid and Harvest Time on Yield and Substrate Composition for Biogas Production. In Proceedings of the Research for Rural Development 2009, Jelgava, Latvia, 22 May 2009; Volume 15, pp. 44–49.

72. Amon, T.; Kryvoruchko, V.; Amon, B.; Zollitsch, W.; Pötsch, E. Biogas production from maize and clover grass estimated with the methane energy value system. In Proceedings of the Conference, Engineering the Future (AgEng'04), Leuven, Belgium, 12–14 September 2004.

73. Qasem, J.R.; Hill, T.A. Growth, development and nutrient accumulation in *Senecio vulgaris* L. and *Chenopodium album* L. *Weed Res.* **1995**, *35*, 187–196. [CrossRef]

74. Gqaza, B.M.; Njume, C.; Goduka, N.I.; George, G. Nutritional assessment of *Chenopodium album* L.(*Imbikicane*) young shoots and mature plant-leaves consumed in the Eastern Cape Province of South Africa. *Int. Proc. Chem. Biol. Environ. Eng.* **2013**, *53*, 97–102.

75. Sarabi, V.; Nassiri-Mahallati, M.; Nezami, A.; Rashed Mohassel, M.H. Effects of common lambsquarters (*Chenopodium album* L.) emergence time and density on growth and competition of maize (*Zea mays* L.). *Aust. J. Crop Sci.* **2013**, *7*, 532–537.

76. Aarssen, L.W.; Hall, I.V.; Jensen, K.I.N. The Biology of Canadian Weeds. 76. *Vicia angustifolia* L., *V. cracca* L., *V. sativa* L., *V. tetrasperma* (L.) Schreb. and *V. villosa* Roth.. *Can. J. Plant Sci.* **1986**, *76*, 711–737. [CrossRef]

77. Lin, T.-Y.; Markhart, A.H. Temperature Effects on Mitochondrial Respiration in *Phaseolus acutffolius* A. Gray and *Phaseolus vulgaris* L. *Plant Physiol.* **1990**, *94*, 54–58. [CrossRef]

78. Lin, T.-Y.; Markhart, A.H. Phaseolus acutifolius A. Gray is More Heat Tolerant than *P. vulgaris* L. in the Absence of Water Stress. *Crop. Sci.* **1996**, *36*. [CrossRef]

79. Hungria, M.; Kaschuk, G. Regulation of N_2 fixation and NO_3^-/NH_4^+ assimilation in nodulated and N-fertilized Phaseolus vulgaris L. exposed to high temperature stress. *Environ. Exp. Bot.* **2014**, *98*, 32–39. [CrossRef]

80. Porch, T.G.; Jahn, M. Effects of high-temperature stress on microsporogenesis in heat-sensitive and heat-tolerant genotypes of Phaseolus vulgaris. *PlantCell Environ.* **2001**, *24*, 723–731. [CrossRef]

81. Gelencsér, T. Mais und Bohnen als Mischkultur. Available online: https://www.bioaktuell.ch/pflanzenbau/ackerbau/mischkulturen/mais-und-bohnen-als-mischkultur.html (accessed on 8 July 2020).

82. Celmeli, T.; Sari, H.; Canci, H.; Sari, D.; Adak, A.; Eker, T.; Toker, C. The nutritional content of common bean (*Phaseolus vulgaris* L.) landraces in comparison to modern varieties. *Agronomy* **2018**, *8*, 166. [CrossRef]

83. MtPleasant, J. Food yields and nutrient analyses of the three sisters: A Haudenosaunee cropping system. *Ethnobiol. Lett.* **2016**, *7*, 87–98. [CrossRef]

Publisher's Note: MDPI stays neutral with regard to jurisdictional claims in published maps and institutional affiliations.

 *agriculture*

Review

Diversity of Species and the Occurrence and Development of a Specialized Pest Population—A Review Article

Anna Wenda-Piesik [1,*] and Dariusz Piesik [2]

[1] Department of Agronomy, UTP University of Science and Technology, 7 Kaliskiego, 85-796 Bydgoszcz, Poland
[2] Department of Biology and Plant Protection, UTP University of Science and Technology, 7 Kaliskiego, 85-796 Bydgoszcz, Poland; piesik@utp.edu.pl
* Correspondence: apiesik@utp.edu.pl

Abstract: The trophic interactions between plants and herbivorous insects are considered to be one of the primary relationships in the occurrence and development of specialized pest populations. Starting from the role of multicropping and the types of mixtures through the ecological benefits of intercropped plants, we explain the ecological conditions that contribute to the occurrence of pest populations. The dynamics of pest populations in crop occur in stages with the survival and development of pest in source of origin, invasion and distribution in crops, development and survival of the population, emigration to the another crop and (or) change of habitat. Possible effects of each stages are described based on the camouflage of visual effects, olfactory effects and reversal of feeding preferences. Fundamental theories of natural enemies and concentration of food resources have been explained to refer to the empirical data.

Keywords: intercropping; mixed crop; herbivores; pest population; natural enemy

Citation: Wenda-Piesik, A.; Piesik, D. Diversity of Species and the Occurrence and Development of a Specialized Pest Population—A Review Article. *Agriculture* 2021, 11, 16. https://doi.org/10.3390/agriculture11010016

Received: 6 December 2020
Accepted: 24 December 2020
Published: 28 December 2020

1. Introduction

Agrosystems provide the food source for the human population which are vulnerable to serious quantitative and qualitative losses due to the occurrence of specialized crop pests [1,2]. Agrocenoses, through their floristic compositions, can regulate the change patterns of diversity and ecological processes between plants and pests through a variety of mechanisms, particularly trophic and behavioral regulation [3–5]. Throughout the world, monoculture (single species) cropping is the most intensive method of plant production. It is the most simplified cultivation method and its aim is to maximize yield and net profit. However, the growth of monoculture system is associated with biological problems: monocultures are more susceptible to pests, diseases and weeds. As monoculture continues, the phytosanitary condition becomes increasingly unstable and requires absolute chemical protection via intensive programs. Pest control in monocultures is based primarily on the use of chemical plant protection products of all generations of pesticides [6,7]. An alternative approach to growing some crop species is inter- and intra-species intercropping. Such crops are subjected to less pest pressure and can therefore be controlled without the intervention of chemical agents [8].

The greater the degree of differentiation in agroecosystems, the more stable the systems that regulate pest populations become when compared with monocultures and productivity is not as compromised [9,10]. A phenomenon that positively influences the efficiency of mixed crops is complementarity [11], due to different species being able to make better use of the habitat's resources, which, in turn, translates into increased plant productivity and total yield [12]. Mixed crops can also better counteract soil erosion and degradation of organic matter, contributing to an increase in the content of organic carbon and nitrogen in soil [13–15].

The trophic interactions between plants and herbivorous insects are considered to be one of the primary relationships that occur in agrocenoses. The presence of pests is regarded as one of the most important biotic factors that affect consecutively cultivated plants during each growing season. The cultivation of only one plant species (especially in monocultures) results in the development of specialized phytophages and, consequently, leads to a reduction in plant productivity. Depending on plant succession, the development of the pest population may be completely or partially limited.

Plant species utilized in crop systems can either improve or worsen the phytosanitary quality of the site for each plant [16]. According to the theory of crop rotation, strategies for the continuation of growth must include preventing so-called crop rotation diseases or, at a minimum, establishing an environment that is not conducive to the excessive development of pest populations [17]. However, crop rotation is not the sole approach that leads to a reduction in populations of pests, pathogens or weeds. Recognizing the crucial role of pest control, researchers are utilizing other methods in the search for new solutions; for instance, resistance breeding or different methods of plant cultivation [18]. In order to reduce the risk of crop failure, which is influenced by the gradations of specialized pest populations, and, at the same time, to ensure crop yield stability, intercropping should be introduced as often as possible into crop production systems [19].

2. Multicropping and Types of Mixtures

According to the literature, multicropping is defined as a practice of consecutively sowing different plants in the same field during a single growing season and many different types of multicropping systems exist. In reference to multicropping systems, Andrews and Kassam [20], Perrin [21] and Willey [22] also include the practice of mixed cropping, which involves planting two or more plant species simultaneously in the same field; these different species coexist either for a limited time or for the whole duration of the growing season. On the other hand, multicropping does not involve the following: single species cultivation (in the same field and for the entire growing season), sowing winter plants in the same growing season subsequent to harvesting spring or winter plants or the cultivation of winter crops in monoculture. It also excludes permanent grassland or sowing perennial monospecies grasses or small-seed legumes on arable land.

Considering the spatial distribution of different plant species and taking into account the length of time that they co-occur, we can distinguish the following types of multi-cropping systems: (1) consecutive crops: during one growing season, two (seldom more) short-term crops, such as mulching crop and spring barley *Hordeum vulgare* L. are sown (in the same field) in successive, relatively short intervals of time; (2) variable crops: a single plant species is introduced into an existing crop of another species. In Poland, this method of cultivation is referred to as "undersowing." The overlap (period of time in which both species coexist) fluctuates from a few to several weeks; for example, seradella *Ornithopus* spp. is introduced into an existing crop of winter rye *Secale cereale* L.; (3) intercrops ("co-crops" or mixed crops): two or more plant species (including varieties of a single species) are cultivated simultaneously. In this instance, the developmental process overlaps in space and time, for example, a mixture of barley and oats *Avena sativa* L.

Intercrop methods can be further divided into the following groups according to their cultivation pattern: (1) rowless plant mixtures: plants of different species (or plant varieties) are sown according to assumed proportions; their placement, however, occurs at random, which results in an unsystematic or mosaic crop pattern, for example, sowing clover seeds with ryegrass *Lolium perenne* L.; (2) row crop mixtures: a mixture of two or more plant species is attained by placing seeds in regular rows but with an irregular distribution pattern within the rows, for example, a mixture of barley and peas *Pisum sativum* L.; (3) inter-row cultivation: plants of each species are alternately arranged and placed in separate, uniform rows, such as single, double or multiple rows. This is a special type of strip cultivation that is used primarily for the cultivation of vegetables. It allows for an independent cultivation technology to be used for individual species; (4) coordinated and rowless mixed cultivation: one species is grown in rows and the field distribution of the remaining species is random; for example, spring barley is sown in rows and alfalfa

Medicago sativa L. is distributed randomly. Therefore, a crop mixture can be defined as the process of simultaneously cultivating two or more species or varieties of arable crops in the same field. Species or varieties, described as mixture components, are usually sown and harvested at the same time. In special cases, however, both the sowing of seeds and the collection of individual species may be performed at different times. 'Simultaneous cultivation' refers to cultivation in one ecological niche for a significant period of the growing season [20]. Furthermore, multicropping systems also include mixtures: spatially arranged crops (where plants of each species are sown in separate rows) and crops characterized by an irregular presence of species within each of the rows [22].

In traditional field crops, both Asian and some tropical regions have the highest share of mixed crops [23]. This particularly applies to such mixtures as coconut *Cocos nucifera* L. and pineapple *Ananas comosus* (L.) Merr, corn *Zea mays* L. and potato *Solanum tuberosum* L., corn and sweet potato *Ipomoea batatas*, sorghum *Sorghum bicolor* (L.) Moench and peas and beans *Phaseolus vulgaris* L. and corn. The achievements of genetics and breeding programs of grasses and clovers have led to the adaptation of varieties of these species for mixed crops for lawn and forage use. Much attention has been paid to research into multi-cultivar crops in common wheat, *Triticum aestivum* L. and rice, *Oryza sativa* L. It is well known that such crops result in greater productivity per acreage because individual cultivars use habitat resources, such as water, light and soil components, more efficiently [24,25]. In countries where agriculture is less developed, traditional crops have always been mixed because of the scarcity of arable land (rarely exceeding 1.5 ha) [26] and this practice has reduced the risk of crop failure [27]. In Central Europe, intercropping involves the utilization of plants from the *Poaceae* and *Papilionaceae* families. The following mixtures are used: mixed cereals of various species, mixtures of varieties of one type of grain (most often barley), mixtures of legume species, mixtures of cereals and legumes, mixtures of small-seed legume species with grasses and mixtures of grass species. The latest research shows that in the cropping of maize with common beans or garden nasturtium *Tropaeolum majus*, yields of dry matter were obtained in comparable quantities and qualities to those resulting from the cultivation of maize alone. This study showed that the intercropping of maize in Central Europe with flowering partners can be a suitable alternative to growing maize alone and can increase field biodiversity [28]. Corn and common beans in co-cultivation is one of the most common food crop production practices in small farms in Sub-Saharan Africa (SSA). In Europe, other forms of multicropping involve introducing undersown, small-seed legumes (or grasses) into cereals and mulching crops, while the strip system is predominantly used in the cultivation of field vegetables. In Poland, the possibilities of their utilization are limited primarily by the length of the growing season. Therefore, given climatic conditions, only certain types of multicropping techniques can be used. Many years of research conducted in Poland have shown that cereal mixtures (especially barley with oats) produced higher yields than pure crops of the same varieties, mainly due to an increase in the leaf area index *LAI* and lend equivalent ratio *LER* [29]. Comparing the organic management system with the integrated management system revealed that the average gross margin (less profit) was twice as high in the mixtures grown in the organic system [30]. However, when deciding to make changes in crop selection, one should take into account the consequences of decreasing the use of mixed crops, as this may hinder the implementation of self-sufficiency and land-use efficiency programs [31].

3. Benefits of Growing Plants in Mixtures

The biodiversity of farmlands has significantly declined, which can be explained by the intensification of agricultural production [32–34]. In consequence, this decline may reduce the abundance of natural enemies and their effects on pest species [35–38].

It is well known that the intensification of agriculture is one of the main causes of biodiversity loss [39] and also has a negative effect on ecosystems [40,41]. Thus, there is a need for more sustainable agricultural practices [42]. Diversification practices (e.g., intercropping or diverse field margins) were intensively used for many centuries and, to date, are well accepted as one of the most promising practices to maintain the biodiversity of ecosystems. Moreover, they may increase productivity in widely utilized agricultural systems [43].

Intercropping plays an important role in controlling many pest species and protecting beneficial insects, which are essential for enhancing biodiversity in an agroecosystem [44–47]. Not surprisingly, it is also important to consider the degree to which host plants are resistant to aphids (*Aphis* spp.). In the intercropping system, wheat cultivars that are resistant to cereal pests may reduce cotton aphids, *Aphis gossypii* L. more effectively than an aphid-resistant variety [48].

Many scientific activities have highlighted the effects of plant diversification on pests, pathogens and beneficial organisms in the agricultural landscape. The results of these studies suggest that habitat manipulation (e.g., intercropping) and rotation can considerably improve both disease and pest management [49].

Intensive agriculture has achieved many advances in agroecosystem productivity. Intensive cropping systems prefer specialized plant group (e.g., cereals) and replace diverse plant ecosystems with monoculture. This not only leads to the loss of cultivated plant resources but also reduces the numerous benefits provided by biodiversity within agroecosystems (e.g., biological control) [50].

The benefits of multicropping for plant cultivation include the development of plant species, for example, an increase in nitrogen uptake by cereals that are cultivated in a mixture with legumes [51]; the efficient use of solar energy in mixtures of monocotyledonous and dicotyledonous plants [52,53]; the minimization of self-poisoning in some crops [54,55]; the incidence of "soil fatigue" [56,57]; a significantly more efficient use of water and nutrients [58]; soil profile; the complementary use of space [54,57,59]; the formation of dense soil cover [60]; the limitation of pests and crop diseases [56,61–63]; the growth of certain species in conditions that are unfavorable for other species [59,64,65]; and higher productivity of multispecies communities as compared with monospecies systems. Mixed cultivation fully supports the various arguments presented in favor of this type of cultivation system [66–69].

4. Ecological Conditions That Contribute to the Occurrence of Pest Populations

The widespread use of chemical plant protection products has caused a number of negative changes and problems, among which an increase in pest resistance and harm to plant pollinators are key effects. In this regard, EU regulations have reduced the spectrum of allowable pesticides and announced a green deal for Europe, according to which the use of pesticides is to be restricted by 30% within 10 years. The use of alternatives to chemical-based methods for pest control, including the increased emphasis on natural enemies, primarily aims to increase the biodiversity of the biocenosis [70,71].

In some integrated pest management systems, the use of mixed crops is a practice to prevent excessive pests. In Mubi, Adamawa and Nigeria, the intercropping of cowpea *Vigna unguiculata* (L.) Walp and sorghum significantly reduced the aphid population (*Aphis craccivora* Koch) compared with the sole crops of these species [72]. However, Oso and Falade [73] stated that intercropping may support other practices but, on its own, may not necessarily resolve increasing pest populations or reduce the pest burden in all situations. Any cropping system with high pest pressure can be managed relatively early as the predator population increases. The start of vegetation growth is always a critical period (autumn in the case of winter crops and spring in the case of spring plants) when the ratio of predator to pest is the highest. It is during this time that the pest population is most likely to be suppressed by predators [74].

Biodiversity is defined as species richness; namely, the variety and variability of species at all trophic levels of any given biocenosis. In complex biocenoses, determining the diversity of animals on a local scale (of any specific ecosystem) depends on the heterogeneity in space, predation and competition. Predation plays a dominant role in shaping the diversity of organisms. In simple biocenoses, however, it is competition that constitutes the most significant factor in organism diversification, which intensifies with the occurrence of highly specialized herbivores that exhibit strong preferences for narrow ecological niches [75]. Diversity in any ecosystem should be treated holistically and it is crucial to understand and carefully consider the role of all trophic levels within it. Among strategies that aim to reduce pesticides, biological pest control is the safest and pro-ecological service for the entire natural environment. However, increasing reservoirs of natural enemies and their population sizes has rarely been the subject of research. The dynamics of the populations of pests and their natural enemies require time and the maintenance of ecological control mechanisms in the agricultural system, which should be studied in the growing cycle and repeated over multiple years [76]. Price et al. [60] emphasized the importance of various interactions between the plant, the herbivorous insect and the herbivore's natural enemies. For example, it is not possible to fully understand the relationship between a plant and its pest without careful consideration of the impact caused by the insect's natural enemies. Consequently, the importance of each trophic level cannot be overlooked.

Barren biocenoses are characterized by relatively small numbers of dominant species and the presence of a substantial quantity of individual species (per unit area). This phenomenon applies to both the producer and consumer levels. These simple systems are significantly more susceptible to an increased presence of a single insect species than any other natural ecosystem [77]. Bey-Bienko [78] provides an example of typical changes in the composition of fauna as a result of a natural system's transformation into arable cropland. The author states that in the natural steppe ecosystem, the number of insect species was 312, while the number of organisms per square meter was about 159. The relationship between the diversity of species and the number of individual species per area unit was inverted after field conversion into a monoculture of wheat. As a result, the number of species dropped to 135, while the number of organisms per square meter increased to 341. Repeated cultivation of the same species in large spaces favors the outbreak of pests. The separation of plants in time (i.e., crop rotation) or in space (multiple crops) can potentially reduce herbivorous insects. In agrocenoses, one approach that results in the differentiation of species or structural differentiation of the canopy involves adding a taxonomically foreign plant to the cultivation of another species or the simultaneous cultivation of genetically diversified plants of the same species. Some authors believe that diversified cultivation requires the presence of undesirable plant species, that is, weeds [79]. Diversified or multispecies cultivation systems contribute to the increased stability of the agrocenosis and, as a practical benefit, the reduction in pest populations [21].

Stability is one of the most important, naturally occurring features of biocenoses. Its disruption or change has a negative effect on the abundance of all populations that exist in the biocenoses. As a result of this stability, it is possible to maintain a relatively consistent influence from disruptive factors [75]. It has been established that the greater the species diversification in any given plant community, the greater the efficiency of the entire trophic network, which affects the balance between the populations of herbivorous and predatory insects [80]. Therefore, an increase in diversity leads to an increase in stability due to properly functioning self-regulating mechanisms of biocenoses. Elton (quoted by Krebs) confirms this thesis by stating: "A sudden explosion in pests' population occurs more often in simple biocenoses or on areas transformed by man" [75]. Because of genetic uniformity and the relatively short period of existence of a given field, agrocenoses are characterized by little biotic diversity and, consequently, limited stability. Therefore, many researchers consider the excessive simplification of agroecosystems to be the primary cause of considerable yield losses [81,82]. It has been estimated that the reduction in global food resources due to pest activity amounts to approximately 13% annually [77].

5. Diversity of Crop Species and the Occurrence of Pests

Herbivorous insects exhibit selective preferences towards host plants. In natural biocenoses, plant communities consist of numerous and unrelated species. Herbivorous insects, when looking for a niche that suits their preferences, are guided by chemical or visual stimulators that emanate from plants. Even insects with a fairly broad foraging spectrum demonstrate food preferences and, therefore, inhabit communities with micro-climates that are most suitable to their needs and requirements [64,82]. In mixed crops, the spatial dispersion of hosts is the main factor that influences the dynamics of the insect population. Table 1 provides examples of pests that have altered their behaviors or the development of their populations as a consequence of diversity in crop species.

Table 1. Examples of crop pests for which changes in the behavior or development of the population have been observed due to intercropping.

Name of Pest and Family	Host Plant	Type of Intercropping	Changes in the Pest Behavior and Pest Population		References
Acalymma vittata Chrysomelidae	Cucumber	Inter-row cultivation, cucumber and corn or broccoli in separate rows	(a) (b) (c)	Three times fewer beetles than in pure cucumber crop Reduction in the reproductive rate Decrease in the period of foraging	[20]
Phyllotreta cruciferae Goeze Chrysomelidae	Broccoli	Inter-row cultivation, broccoli in rows and white clover between rows	(a) (b)	Colonization of broccoli beetle populations is 1.3 times slower than in the pure stand of broccoli Two-fold increase in the migration time of beetles to other crops	[61]
Phyllotreta cruciferae Goeze Chrysomelidae	Broccoli	Inter-row cultivation, broccoli in rows and vetch and bean between rows	(a) (b) (c)	Decrease in the foraging period Abandonment of mixed crops Decreasing of the population	[83]
Phyllotreta cruciferae Goeze Chrysomelidae	Cabbage	Row-crop mixture, cabbage and tobacco or tomato in separate rows	(a) (b)	Significant reduction in the pest population of coordinate-mixed cultivation with shortening of beetle feeding time More than 3 times fewer second-generation beetles as compared with the cultivation in the pure stand	[82]
Aphis craccivora Koch Aphididae	Groundnut	Row-crop mixture, groundnut and common beans in separate rows	(a) (b)	Common bean's sticky tendrils kept aphids away A reduction in aphids as vectors resulted in a decrease in the virus that causes rosette disease of groundnut	[84]
Oulema spp., Chrysomelidae	Oat, barley	Row-crop mixture of both cereals	(a)	Mixed cultivation of each species reduces the degree of damage to oat leaves by 48% and barley by 51% compared with pure stands	[85]
Rhopalosiphum padi L., *Sitobion avenae* L. Aphididae	Barley	Row-crop mixture of barley with yellow lupine and pea	(a)	The number of aphids on barley heads was 3–6 times lower in crops with legumes	[86]

Perrin lists four aspects that determine the development of the pest population in mixed cultivation: the infestation of the crop by the pest (colonization), the development of its population, the dispersion of herbivores in the cultivation and the presence of natural pests [20]. The individual stages of development of the pest population and the possible effects of changes in insects are presented in Figure 1.

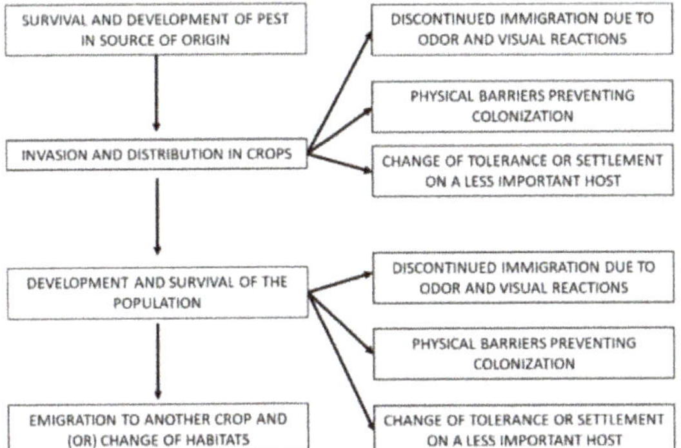

Figure 1. Stages in the dynamics of pest populations resulting from mixed cultivation. Possible effects are listed on the right [87].

6. Colonization of Crops by Pests

The following factors influence the colonization of mixed crops by specialized pests:

(1) Camouflage of visual effects. A mixed crop becomes visually unattractive to incoming pests since host plants are often obscured by non-host plants with longer shoots or branch shapes. Consequently, insects' perception of the entire cultivation area becomes skewed. "Foreign" plants constitute a physical barrier to the spread of pests; they also function as "traps" [83,87].

(2) Olfactory (aromatic) effects. Attractants or feeding stimulants secreted by the host plants play a significant role and determine the way herbivores orient themselves in the environment. Strongly aromatic plants such as tomato *Solanum lycopersicum* L., garlic *Allium sativum* L., onion *Allium cepa* L. and tobacco *Nicotiana tabacum* L. (while cultivated together with other species) may disturb the olfactory perception of the habitat [82,84,88].

(3) Reversal of feeding preferences. In some cases, pests show strong preferences for and inhabit only certain plant species cultivated in a particular mixture. As a result, pests become "distracted," which, in turn, ensures the protection of other, more valuable plant species. Utilization of this phenomenon is exemplified by the planting of alfalfa on the perimeter of cotton *Gossypium* spp. L. crops in California. Cotton bugs *Lygus Hesperus* L. cause significant damage to cotton plantations. However, their adverse effects on cotton fields are reduced considerably due to the insects' feeding preferences and their apparent attraction to alfalfa [21]. Without the introduction of alfalfa, the insects' impact would be significantly more pronounced.

The above-mentioned factors for reducing the insect colonization of mixed crops are of particular importance in the case of populations of mobile pests that inhabit crops at the beginning of each growing season (e.g., winter beetles looking for complementary feeding crops).

7. Development of the Pest Population in Mixtures

A sudden increase in the population of pests occurs when individual species are easily capable of locating food, shelter and favorable conditions for reproduction [89]. Because mixtures reduce the population of host plants, they change the design and physiognomy of the cultivation and have a negative effect on the microclimate of specialized pest species. Farell observed that the sticky leaf tendrils of the common bean were capable of "catching" the aphid *Aphis craccivora* Koch a vector of the peanut virus (peanut mottle virus, peanut stripe virus and peanut stunt virus), thus effectively limiting the development of the insect population [84]. The benefits of mixed cultivation are contingent upon the time of insect emergence in relation to the stage of the plant developmental process. The negative effects of pests are more pronounced during the most critical stages of plant development since plants are most susceptible to damage during their emergence, as well as during the flowering process [21].

8. Pest Distribution in Cultivation

Inhibiting the spread of the pest population is possible when host and non-host plants grow together in a particularly unfavorable system for herbivores. The scattering of the cabbage flea *Phyllotreta cruciferae* Goez9e was inhibited on cabbage *Brassica oleracea* L. var. *capitata* L. when cabbage was grown in a row around the perimeter of a meadow to a much greater extent than in a plot in the same meadow consisting of several rows that were only 45 cm apart [90]. It was found that the same pest in a mixture of broccoli *Brassica oleracea* L. var. *italica* Plenck and vetch *Vicia sativa* L. or field bean *Vicia faba* L. var. *minor* Peterm. wasted considerable time and energy on disentangling from vetch shoots and finding the right host among horse bean plants, which resulted in a rapid reduction in its population [83]. The decreased availability of the niche and large distances between plants reduce the relative quality of the insect environment, which, in turn, may lead to the emigration of pests to other, more attractive crops. This effect, in which one plant helps another plant to defend itself effectively against pests, is called "companion immunity." Examples of crops with this type of resistance are listed in Table 1.

9. The Role of Natural Enemies

The relationship between a plant and an insect cannot be considered without taking into account the third trophic level: natural enemies, which are considered plant allies [77]. The more diversified the cultivation, the greater the variety and abundance of herbivorous predators and parasitoids. Therefore, the simultaneous cultivation of several species may alleviate and/or stabilize the relationship between a pest and its natural enemy [91]. Long-term crops are of particular importance here, since the stability of the relationship between plant, phytophage and entomophage is positively influenced by an extended period of time [21].

Grape phylloxera, *Viteus vitifoliae* (Fitch), is regarded as the most economically important pest worldwide for commercial grapevines (*Vitis* spp.). Grape phylloxera causes the most economic damage in its root-feeding stages as compared with leaf-feeding stages [92]. Research on grape phylloxera has been extensive because this pest ravaged European vineyards and most of the basic work on phylloxera biology and control was carried out prior to 1920. Granett et al. [93] summarized the major constraints that explain why chemical control has been inefficient in root-galling grape. No efficient biological control method has been developed to date, though many general natural enemies of phylloxera exist [94]. An organic management strategy could reduce root necrosis but it produces no effect on the number of phylloxerae: this observation may be due to the microbial ecology and soil suppression of pathogens [95]. Soil type may also influence phylloxera survival and its spread [96]. However, control methods that may be efficient and practical for supporting successful pest control remain unclear and require more testing.

The effects of grape–tobacco intercropping on populations of grape phylloxera were evaluated in a field in which egg and nymph mortality and female fecundity were significantly affected. It was reported that grape phylloxera populations in the intercropping systems were lower compared with the monoculture pattern and they decreased each year. Vine trees were in better condition upon continuous intercropping with tobacco [97]. Intercropping is also effective in reducing mantis cruciferous *Plutella xylostella* L. populations but the underlying mechanisms are elusive [98]. For example, when exposed to three different types of host plants (*Brassica campestris* L., *B. juncea* Coss. and *B. oleracea* L.), the flight frequency of adult *P. xylostella* females increases, while its fecundity is weakened [99]. Many researchers have studied the positive or negative impact of the infestation of pest species [47,100,101]. Bregante and Matta [102] studied the intercropping of corn and bean and Omar et al. [103] conducted field trials to study the effect of intercropping cotton and cowpea on the populations of aphids, whitefly and bollworm. Ma et al. [104] examined the strip cropping of wheat and alfalfa to improve the biological control of the cereal aphid, *Sitobion avenae* (Fabr.) by the mite, *Allothrombium* Berlese (Acari: Trombidiidae). It is well documented that wheat–garlic intercropping can reduce the population of *S. avenae* by promoting natural enemies [105]. Similar studies have also been performed in wheat and oilseed rape, *Brassica napus* L. [106], cowpea and sorghum [72] and wheat and pea [107,108].

10. Theories of Natural Enemies and Concentration of Food Resources

During the examination of pure as well as mixed cabbage crops, Root [90] observed that the number of pests and their average biomass per 100 g of consumed food was always higher in pure cabbage sowing. In order to explain this phenomenon, the author presented two hypotheses. The natural enemies hypothesis attributes the lower pest density to a more diversified environment, where higher numbers of predator species and insect parasitoids are present and the abundance of their populations is increased [77,109]. Proponents of this theory regard the enemies of natural insects as the main factor in regulating populations of pests. An alternative hypothesis is derived from the theory of food resource concentration. In non-uniform, short-term crops, the effectiveness of natural enemies in reducing phytophages may not be as effective as the mere fact of decreased food concentration. In mixed sowing, specialized pests are deprived of a sufficient food supply, proper breeding base and adequate shelter. Therefore, they show a distinct preference towards single-species compact sowing, where the concentration of host plants is sufficient to maintain all necessary vital functions [90]. Most researchers strongly support the hypothesis of the concentration of food resources [61,79,81–83,87].

On the other hand, opponents argue that the two general theories that explain the interaction between an insect and cultivation in a multiple-plant system cannot be applied to individual pests and their populations. Speaking against the hypothesis of resource concentration, Helenius [110] gives the example of the cereal aphid, *Rhopalosiphum padi* L. in a mixture of oats and field beans. The substantial abundance of the oat plants resulted in a greater density of aphids due to the more pronounced aggregation of colonies established by re-emigrants on a single plant. The activity of natural enemies may also decrease in crops with a variety of species, especially if they become attracted by specific visual or olfactory stimulants, the reception of which may be disturbed by "concealment" by other plant species. Smith [111] postulates that this mechanism causes a disruption in the proper perception of the habitat by the infiltrating herbivorous insects. Moreover, increasing crop biodiversity, such as by strip intercropping, can promote biological pest control in agroecosystems [74,112].

11. Conclusions

The spatial and/or temporal separation of host plants is contingent upon the behavior and development of herbivorous insects. Reducing pest populations can be realized by recognizing and identifying their feeding preferences. The more pronounced the feeding preferences, the greater the reduction in the population. Consequently, the damage to host plants grown in mixed sowing systems will be considerably reduced. Monophagous insects are specific in this regard. The slight alteration of a host plant's canopy renders monophagous insects unable to locate an adequate food supply and to establish a suitable breeding base. A significant reduction in the population of oligophagous insects (insects whose host spectrum is in the botanical family) is expected to occur in mixtures of adequately spaced botanical taxa, for example, damage to cereal plants can be reduced by introducing the cereal leaf beetle to cereal–legume mixtures. Numerous empirical data and some theoretical considerations suggest that, in mixed crop systems, the reduction in pest populations is predominantly linked to the availability of food sources and less so to an impact or threat posed by their natural enemies.

Author Contributions: Conceptualization, A.W.-P. and D.P.; formal analysis, A.W.-P. and D.P.; investigation, A.W.-P.; resources, D.P.; writing—original draft preparation, A.W.-P.; A.W.-P. and D.P. All authors have read and agreed to the published version of the manuscript.

Funding: This research received no external funding.

Institutional Review Board Statement: Not applicable.

Informed Consent Statement: Not applicable.

Conflicts of Interest: The authors declare no conflict of interest.

References

1. Garcia, L.; Celette, F.; Gary, C.; Ripoche, A.; Valdes-Gomez, H.; Metay, A. Management of service crops for the provision of ecosystem services in vineyards: A review. *Agric. Ecosyst. Environ.* **2018**, *251*, 158–170. [CrossRef]
2. Godfray, H.C.J.; Beddington, J.R.; Crute, I.R.; Haddad, L.; Lawrence, D.; Muir, J.F.; Pretty, J.; Robinson, S.; Thomas, S.M.; Toulmin, C. Food security: The challenge of feeding 9 billion people. *Science* **2010**, *327*, 812–818. [CrossRef] [PubMed]
3. Karp, D.S.; Chaplinkramer, R.; Meehan, T.D.; Martin, E.A.; Declerck, F.; Grab, H.; Gratton, C.; Hunt, L.; Larsen, A.E.; Martínezsalinas, A. Crop pests and predators exhibit inconsistent responses to surrounding landscape composition. *Animal* **2018**, *115*, 7863–7870. [CrossRef] [PubMed]
4. Snyder, W.E. Give predators a complement: Conserving natural enemy biodiversity to improve biocontrol. *Biol. Control* **2019**, *135*, 73–82. [CrossRef]
5. Wan, N.F.; Ji, X.Y.; Deng, J.Y.; Kiaer, L.P.; Cai, Y.M.; Jiang, J.X. Plant diversification promotes biocontrol services in peach orchards by shaping the ecological niches of insect herbivores and their natural enemies. *Ecol. Indic.* **2019**, *99*, 387–392. [CrossRef]
6. Zhao, Z.H.; Hui, C.; Hardev, S.; Ouyang, F.; Dong, Z.; Ge, F. Responses of cereal aphids and their parasitic wasps to landscape complexity. *J. Econ. Entomol.* **2014**, *107*, 630–637. [CrossRef]
7. Ratnadass, A.; Paula, F.; Jacques, A.; Robert, H. Plant species diversity for sustainable management of crop pests and diseases in agroecosystems: A review. *Agron. Sustain. Dev.* **2012**, *32*, 273–303. [CrossRef]
8. Lai, R.Q.; You, M.S.; Jiang, L.C.; Lai, B.T.; Chen, S.H.; Zeng, W.L.; Jiang, D.B. Evaluation of garlic intercropping for enhancing the biological control of *Ralstonia solanacearum* in flue-cured tobacco fields. *Biocontr. Sci. Tech.* **2011**, *21*, 755–764. [CrossRef]
9. Lai, R.Q.; You, M.S.; Lotz, L.A.P.; Vasseur, L. Responses of green peach aphids and other arthropods to garlic intercropped with tobacco. *Agron. J.* **2011**, *103*, 856–863. [CrossRef]
10. Lai, R.Q.; Zeng, W.L.; Jiang, G.H.; Li, L.Y.; Xie, X.H. Preliminary studyoncontrol effects of ethanol extracts from garlic plant against *Ralstonia solanacearum* and TMV. *J. Yunnan Agric. Univ.* **2011**, *26*, 284–287.
11. Honga, Y.; Berentsen, P.; Heerink, N.; Shi, M.; van der Werfe, W. The future of intercropping under growing resource scarcity and declining grainprices—A model analysis based on a case study in Northwest China. *Agric. Syst.* **2019**, *176*, 102661. [CrossRef]
12. Yu, Y.; Stomph, T.J.; Makowski, D.; van der Werf, W. Temporalniche differentiation increases the land equivalent ratio of annual intercrops: A meta-analysis. *Field Crop Res.* **2015**, *184*, 133–144. [CrossRef]
13. Zhang, C.C.; Dong, Y.; Tang, L.; Zheng, Y.; Makowski, D.; Yu, Y.; Zhang, F.S.; van der Werf, W. Intercropping cereals with faba bean reduces plant disease incidence regardless of fertilizer input; a meta-analysis. *Eur. J. Plant Pathol.* **2019**, *154*, 931–942. [CrossRef]
14. Feike, T.; Doluschitz, R.; Chen, Q.; Graeff-Hönninger, S.; Claupein, W. How to overcome the slow death of intercropping in the North Chinaplain. *Sustainability* **2012**, *4*, 2550–2565. [CrossRef]
15. Cong, W.; Hoffland, E.; Li, L.; Six, J.; Sun, J.; Bao, X.; Zhang, F.; van der Werf, W. Intercropping enhances soil carbon and nitrogen. *Glob. Change Biol.* **2015**, *21*, 1715–1726. [CrossRef]
16. Gonet, I.; Gonet, Z. Effect of the frequency of spring barley or oat cultivation in the same field on the grain yield and the presence of the cereal nematode (*Heterodera avenae*) in the soil. *Pam. Puł.* **1982**, *77*, 49–62. (In Polish)
17. Niewiadomski, W. The science of crop rotation—Status and prospects. *Post. Nauk Roln.* **1995**, *3*, 127–139. (In Polish)
18. Gacek, E. The use of plant genetic diversity in the control of crop plant diseases. *Post. Nauk Roln.* **2000**, *5*, 17–25. (In Polish)
19. Raseduzzaman, M.; Jensen, E.S. Does intercropping enhance yield stability in arable crop production? A meta-analysis. *Eur. J. Agron.* **2017**, *91*, 25–33. [CrossRef]
20. Andrews, D.J.; Kassam, A.H. The importance of multiple cropping in increasing world food supplies. *Am. Soc. Agron.* **1976**, *27*, 32.
21. Perrin, R.M. Pest management in multiple cropping systems. *Agro-Ecosyst.* **1977**, *3*, 93–118. [CrossRef]
22. Willey, R.W. Intercropping—Its importance and research needs. Competition and yield advantages. *Field Crop Abstr.* **1979**, *32*, 1–10.
23. Geno, L.; Geno, B. *Polyculture Production: Principles, Benefits and Risks of Multiple Cropping Land Management Systems for Australia*; Rural Industries Research and Development Corporation: Barton, Australia, 2001.
24. Pretty, J.; Bharucha, Z.P. Sustainable intensification in agricultural systems. *Ann. Bot.* **2014**, *114*, 1571–1596. [CrossRef] [PubMed]
25. Brooker, R.W.; Bennett, A.E.; Cong, W.-F.; Daniell, T.J.; George, T.S.; Hallett, P.D.; Hawes, C.; Iannetta, P.P.M.; Jones, H.G.; Karley, A.J.; et al. Improving intercropping: A synthesis of research in agronomy, plant physiology and ecology. *New Phytol.* **2015**, *206*, 107–117. [CrossRef]

26. Lunze, L.; Abang, M.M.; Buruchara, R.; Ugen, M.A.; Nabahungu, N.L.; Rachier, G.O.; Ngongo, M.; Rao, I. *Integrated Soil Fertility Management in Bean-Based Cropping Systems of Eastern, Central and Southern Africa, Soil Fertility Improvement and Integrated Nutrient Management—A Global Perspective*; Whalen, J., Ed.; InTech: Vienna, Austria, 2012; ISBN 978-953-307-945-5.

27. Giller, K.E. *Nitrogen Fixation in Tropical Cropping Systems*, 2nd ed.; CAB International: Wallingford, UK, 2001.

28. Schulz, V.; Schumann, C.; Weisenburger, S.; Müller-Lindenlauf, M.; Stolzenburg, K.; Möller, K. Row-Intercropping Maize (*Zea mays* L.) with biodiversity-enhancing flowering-partners—Effect on plant growth, silage yield, and composition of harvest material. *Agriculture* **2020**, *10*, 524. [CrossRef]

29. Gałęzewski, L.; Jaskulska, I.; Wilczewski, E.; Wenda-Piesik, A. Response of yellow lupine to the proximity of other plants and unplanted path in strip intercropping. *Agriculture* **2020**, *10*, 285. [CrossRef]

30. Klima, K.; Synowiec, A.; Puła, J.; Chowaniak, M.; Pużyńska, K.; Gala-Czekaj, D.; Kliszcz, A.; Galbas, P.; Jop, B.; Dąbkowska, T.; et al. Long-Term Productive, Competitive, and Economic Aspects of Spring Cereal Mixtures in Integrated and Organic Crop Rotations. *Agriculture* **2020**, *10*, 231. [CrossRef]

31. Zuo, L.; Wang, X.; Zhang, Z.; Zhao, X.; Liu, F.; Yi, L.; Liu, B. Developing grain production policy in terms of multiple cropping systems in China. *Land Use Policy* **2014**, *40*, 140–146. [CrossRef]

32. Rayl, R.J.; Shields, M.W.; Tiwari, S.; Wratten, S.D. Conservation biological control of insect pests. In *Sustainable Agriculture Reviews 28: Ecology for Agriculture*; Gaba, S., Smith, B., Lichtfouse, E., Eds.; Springer International Publishing: Cham, Switzerland, 2018; pp. 103–124.

33. Tomasetto, F.; Tylianakis, J.M.; Reale, M.; Wratten, S.; Goldson, S.L. Intensified agriculture favors evolved resistance to biological control. *Proc. Natl. Acad. Sci. USA* **2017**, *114*, 3885–3890. [CrossRef]

34. Tschumi, M.; Ekroos, J.; Hjort, C.; Smith, H.G.; Birkhofer, K. Predation-mediated ecosystem services and disservices in agricultural landscapes. *Ecol. Appl.* **2018**, *28*, 2109–2118. [CrossRef]

35. Gurr, G.M.; Wratten, S.D.; Landis, D.A.; You, M. Habitat management to suppress pest populations: Progress and prospects. *Annu. Rev. Entomol.* **2017**, *62*, 91–109. [CrossRef] [PubMed]

36. Jacobsen, S.K.; Moraes, G.J.; Sørensen, H.; Sigsgaard, L. Organic cropping practice decreases pest abundance and positively influences predator-prey interactions. *Agric. Ecosyst. Environ.* **2019**, *272*, 1–9. [CrossRef]

37. Muneret, L.; Auriol, A.; Thiéry, D.; Rusch, A. Organic farming at local and landscape scales fosters biological pest control in vineyards. *Ecol. Appl.* **2018**, *29*, 01818. [CrossRef] [PubMed]

38. Shapira, I.; Gavish-Regev, E.; Sharon, R.; Harari, A.R.; Kishinevsky, M.; Keasar, T. Habitat use by crop pests and natural enemies in a Mediterranean vineyard agroecosystem. *Agric. Ecosyst. Environ.* **2018**, *267*, 109–118. [CrossRef]

39. Gamez-Virues, S.; Perovic, D.J.; Gossner, M.M.; Beorschig, C.; Blüthgen, N.; de Jong, H.; Simons, N.K.; Klein, A.-M.; Krauss, J.; Maier, G.; et al. Landscape simplification filters species traits and drives biotic homogenization. *Nat. Commun.* **2015**, *6*, 8568. [CrossRef] [PubMed]

40. Goulson, D.; Nicholls, E.; Botías, C.; Rotheray, E.L. Bee declines driven by combined stress from parasites, pesticides, and lack of flowers. *Science* **2015**, *80*, 1–16. [CrossRef] [PubMed]

41. Rusch, A.; Chaplin-Kramer, R.; Gardiner, M.M.; Hawro, V.; Holland, J.; Landis, D.; Thies, C.; Tscharntke, T.; Weisser, W.W.; Winqvist, C.; et al. Agricultural landscape simplification reduces natural pest control: A quantitative synthesis. *Agric. Ecosyst. Environ.* **2016**, *221*, 198–204. [CrossRef]

42. Bommarco, R.; Kleijn, D.; Potts, S.G. Ecological intensification: Harness in gecosystem services for food security. *Trends Ecol. Evol.* **2013**, *28*, 230–238. [CrossRef]

43. Lithourgidis, A.S.; Dordas, C.A.; Damalas, C.A.; Vlachostergios, D.N. Annual intercrops: An alternative pathway for sustainable agriculture. *Aust. J. Crop Sci.* **2011**, *5*, 396–410.

44. Eskandari, H.; Ghanbari, A. Effect of different planting pattern of wheat (*Triticum aestivum*) and bean (*Vicia faba*) on grain yield, dry matter production and weed biomass. *Not. Sci. Biol.* **2010**, *2*, 111–115. [CrossRef]

45. Konar, A.; Singh, N.J.; Paul, R. Influence of intercropping on population dynamics of major insect pests and vectors of potato. *J. Entomol. Res.* **2010**, *34*, 151–154.

46. Suresh, R.; Sunder, S.; Pramod, M. Effect of intercrops on the temporal parasitization of *Helicoverpa armigera* (Hub.) by larval parasitoid, *Campoletis chlorideae* Uchida in tomato. *Environ. Ecol.* **2010**, *28*, 24852489.

47. Vaiyapuri, K.; Amanullah, M.M. Pest incidence and yield as influenced by intercropping unconventional green manures in cotton. *Madras Agric. J.* **2010**, *97*, 51–57.

48. Ma, X.M.; Liu, X.X.; Zhang, Q.W.; Zhao, J.Z.; Cai, Q.N.; Ma, Y.A.; Chen, D.M. Assessment of cotton aphids, Aphis gossypii, and their natural enemies on aphid-resistant and aphid-susceptible wheat varieties in a wheat-cotton relay intercropping system. *Entomol. Exp. Appl.* **2006**, *121*, 235–241. [CrossRef]

49. Iqbal, N.; Hussain, S.; Ahmed, Z.; Yang, F.; Wang, X.; Liu, W. Comparative analysis of maize-soybean strip intercropping system: A review. *Plant Prod. Sci.* **2018**, *22*, 131–142. [CrossRef]

50. Nassary, E.K.; Baijukya, F.; Ndakidemi, P.A. Productivity of intercropping with maize and common bean over five cropping seasons on smallholder farms of Tanzania. *Eur. J. Agron.* **2020**, *113*, 125964. [CrossRef]

51. Kapoor, P.; Ramakrishnan, P.S. Studies on crop-legume behaviour in pure and mixed stands. *Agro Ecosyst.* **1975**, *2*, 61–74. [CrossRef]

52. Aldrich, R.J. *Ecology of Weeds in Crops. Basics of Weed Control*; Society of Ecological Chemistry and Engineering: Opole, Poland, 1997; p. 461.
53. Hayden, Z.D.; Brainard, D.C.; Henshaw, B.; Ngouajio, M. Winter annual weed suppression in rye–vetch cover crop mixtures. *Weed Technol.* **2012**, *26*, 818–825. [CrossRef]
54. Probst, G. *Field Cultivation Plants. A Textbook of Organic Farming*; Siebeneicher, G.E., Ed.; PWN: Warsaw, Poland, 1997; pp. 136–230. (In Polish)
55. Wien, H.C.; Littleton, E.J. Grain legume improvement program, physiology subprogram. In Proceedings of the IITA Collaborators Meeting on Grain Legume Improvement, Ibadan, Nigeria, 9 June 1975; Luse, A., Rachie, K.O., Eds.; International Institution of Tropical Agriculture: Ibandan, Nigeria, 1975; pp. 127–129.
56. Fularowa, K. Results of experiments with mixed sowing of oats with barley. *Post. Nauk Roln.* **1967**, *5*, 43–50. (In Polish)
57. Trenbath, B.R. Biomass productivity of mixtures. *Adv. Agron.* **1974**, *26*, 177–210.
58. Martin, M.P.L.D.; Snaydon, R.W. Intercropping barley and beans I. Effects of planting pattern. *Expl. Agric.* **1982**, *18*, 139–148. [CrossRef]
59. Gacek, E. The use of biological mechanisms for the prevention of infectious cereal diseases in mixed crops. In Proceedings of the Materials Biotic Arable Environment and the Disease Risk of Plants, Olsztyn, Poland, 7–9 September 1993; pp. 59–65.
60. Price, P.W.; Bouton, C.E.; Gross, P.; McPheron, B.A.; Thompson, J.N.; Weis, A.E. Interactions among three trophic levels: Influence of plants on interactions between insect herbivores and natural enemies. *Ann. Rev. Ecol. Syst.* **1980**, *11*, 41–65. [CrossRef]
61. Elmstrom, K.M.; Andow, D.A.; Barclay, W.W. Flea beetle movement in a broccoli monoculture and diculture. *Environ. Entomol.* **1988**, *17*, 299–305. [CrossRef]
62. Wolfe, M.S. The current status and prospects of multiline cultivars and variety mixtures for disease resistance. *Ann. Rev. Phytopathol.* **1985**, *23*, 251–273. [CrossRef]
63. Wolfe, M.S. Intra-crop diversification: Disease, yield and quality. Crop Protection in Organic and Low Input Agriculture. *BCPC Monogr.* **1990**, *45*, 105–114.
64. Helenius, J. Plant size, nutrient composition and biomass productivity of oats and faba bean in intercropping, and the effect of controlling *Rhopalosiphum padi* (Homoptera: Aphididae) on these properties. *J. Agric. Sci. Fin.* **1989**, *60*, 1–20. [CrossRef]
65. Helenius, J.; Ronni, P. Yield, its components and pest incidence in mixed intercop ping of oats (*Avena sativa*) and field beans (*Vicia faba*). *J. Agric. Sci. Fin.* **1989**, *61*, 15–31.
66. Budzyński, W.; Majkowski, K.; Nikiel, A.; Wróbel, E. Effect of sowing method on the yield of spring barley and oats on the soil of a very good rye complex. *Zesz. Nauk. ART. Olsztyn Roln.* **1980**, *30*, 181–189. (In Polish)
67. Majkowski, K.; Szempliński, W.; Budzyński, W.; Wróbel, E.; Dubis, B. Cultivation of inter-varietal and interspecific mixtures of spring barley and oats. *Rocz. AR Poznań CCXLIII* **1993**, *1*, 85–96. (In Polish)
68. Rudnicki, F.; Wasilewski, P. Comparison of the yield of spring cereal mixtures on the soil of the good rye complex. In Proceedings of the Condition and Prospects for the Cultivation of cereal Mixtures in Poland (in Polish), Poznan, Poland, 2 December 1994; pp. 83–87.
69. Wanic, M. A mixture of spring barley with oats and single-species crops of these cereals in crop rotation. Habilitation dissertation. *Acta Acad. Agricult. Tech. Olst. Agric.* **1997**, *64D*, 50. (In Polish)
70. Saied, A.A.A.; Shahien, F.A.; Hamid, A.M.; El-Zahi, E.S. Effect of certain natural and specific materials on some sucking pests and their associated natural enemies in cotton crop. In Proceedings of the 2nd International Conference in Plant Protection Research Institute, Cairo, Egypt, 21–24 December 2002.
71. Mahmoud, F.M.; Osman, M.A.M. Relative toxicity of some bio-rational insecticides to second instar larvae and adults of onion thrips (*Thrips tabaci* Lind) and their predator *Orius albidipennis* under laboratory and field conditions. *J. Plant Protect. Res.* **2007**, *47*, 391–400.
72. Hassan, S. Effect of variety and intercropping on two major cowpea [*Vigna unguiculata* (L.) Walp] field pests in Mubi, Adamawa State, Nigeria. *J. Horticult. Forest.* **2009**, *1*, 014–016.
73. Oso, A.A.; Falade, M.J. Effect of variety and spatial arrangement on pest incidence, damage and subsequent yield of cowpea in a cowpea/maize intercrop. *World J. Agric. Sci.* **2010**, *6*, 274–276.
74. Liu, J.-L.; Ren, W.; Zhao, W.-Z.; Li, F.-R. Cropping systems alter the biodiversity of ground- and soil-dwelling herbivorous and predatory arthropods in a desert agroecosystem: Implications for pest biocontrol. *Agric. Ecosyst. Environ.* **2018**, *266*, 109–121. [CrossRef]
75. Krebs, C.J. *Organization of Biocenosis II: Predation and Competition in Sustainable Biocenoses*; Ekologia. PWN: Warsaw, Poland, 1996; pp. 471–520. (In Polish)
76. Ouyanga, F.; Su, W.; Zhang, Y.; Liu, X.; Su, J.; Zhang, Q.; Men, X.; Ju, Q.; Ge, F. Ecological control service of the predatory natural enemy and its maintaining mechanism in rotation-intercropping ecosystem via wheatmaize-cotton. *Agri. Ecosyst. Environ.* **2020**, *301*, 107024. [CrossRef]
77. Pimentel, D. Diversification of biological control strategies in agriculture. *Crop Prot.* **1991**, *10*, 243–253. [CrossRef]
78. Bey-Bienko, G.Y. On some regularities in the changes of the invertebrate fauna during the utilisation of virgin steppe. *Rev. Entomol. URSS* **1961**, *40*, 763–775.
79. Risch, S.J.; Andow, D.; Altieri, M.A. Agroecosystem diversity and pest control: Data, tentative conclusions and new research directions. *Environ. Entomol.* **1983**, *12*, 625–629. [CrossRef]

80. MacArthur, R.H. Fluctuations of animal populations and a measure of community stability. *Ecology* **1955**, *36*, 533–536. [CrossRef]
81. Bach, C.E. Effects of plant diversity and time of colonisation on a herbivore-plant interaction. *Oecologia* **1980**, *44*, 319–326. [CrossRef]
82. Tahvanainen, J.O.; Root, R.B. The influence of vegetational diversity on the population ecology of a specialized herbivore, *Phyllotreta cruciferae* (Coleoptera: Chrysomelidae). *Oecologia* **1972**, *10*, 321–346. [CrossRef] [PubMed]
83. Garcia, M.A.; Altieri, M.A. Explaining differences in flea beetle *Phyllotreta cruciferae* Goeze densities in simple and mixed broccoli cropping systems as a function of individual behaviour. *Entomol. Exp. Appl.* **1992**, *62*, 201–209. [CrossRef]
84. Farrell, J.A.K. Effects of intersowing with beans on the spread groundnut rosette virus by *Aphis craccivora* Koch (Hemiptera: Aphididae) in Malawi. *Bull. Entomol. Res.* **1976**, *66*, 331–333. [CrossRef]
85. Wenda-Piesik, A.; Piesik, D. The spring cereals food preferences of *Oulema* spp. in pure and mixed crops. *Elec. J. Pol. Agric. Univ. Agron.* **1998**, *1*, 1–12.
86. Wenda-Piesik, A. Health status of spring barley grown in monocrop and in mixtures with cereals or leguminous plants. *J. Plant Prot. Res.* **2001**, *41*, 388–394.
87. Perrin, R.M.; Phillips, M.L. Some effects of mixed cropping on the population dynamics of insect pests. *Entomol. Exp. Appl.* **1978**, *24*, 385–393. [CrossRef]
88. Aiyer, A.K.Y.N. Mixed cropping in India. *Indian J. Agric. Sci.* **1949**, *19*, 439–543.
89. Andrews, D.J. Intercropping with guinea corn, a biological co-operative: Part I. *Samaru Agric. Newsl.* **1972**, *14*, 20–22.
90. Root, R.B. Organization of a plant-arthropod association in simple and diverse habitats: The fauna of collards (*Brassica oleracea*). *Ecol. Monogr.* **1972**, *43*, 95–124. [CrossRef]
91. Price, P.; Waldbauer, G.P. Ecological aspects of insect pest management. In *Introduction to Insect Pest Management*; Wiley: New York, NY, USA, 1975; pp. 36–73.
92. Ma, K.Z.; Hao, S.G.; Zhao, H.Y.; Kang, L. Strip cropping wheat and alfalfa to improve the biological control of the wheat aphid *Macrosiphum avenae* by the mite *Allothrombium ovatum*. *Agric. Ecosyst. Environ.* **2007**, *119*, 49–52. [CrossRef]
93. Wang, W.L.; Liu, Y.; Ji, X.L.; Wang, G.; Zhou, H.B. Effects of wheat-oilseed rape intercropping or wheat-garlic intercropping on population dynamics of *Sitobion avenae* and its main natural enemies. *Chin. J. Appl. Ecol.* **2008**, *19*, 1331–1336.
94. Wang, W.L.; Liu, Y.; Chen, J.L.; Ji, X.L.; Zhou, H.B.; Wang, G. Impact of intercropping aphid-resistant wheat cultivars with oilseed rape on wheat aphid (*Sitobion avenae*) and its natural enemies. *Acta Ecol. Sin.* **2009**, *29*, 186–191. [CrossRef]
95. Zhou, H.B.; Chen, L.; Chen, J.L.; Liu, Y.; Cheng, D.F.; Sun, J.R. The effect of intercropping between wheat and pea on spatial distribution of *Sitobion avenae* based on GIS. *Sci. Agr. Sin.* **2009**, *42*, 3904–3913.
96. Zhou, H.B.; Chen, J.L.; Cheng, D.F.; Liu, Y.; Sun, J.R. Effects of wheat-pea intercropping on *Sitobion avenae* and the functional groups of its main natural enemies. *Acta Entomol. Sin.* **2009**, *52*, 775–782.
97. Omar, H.I.H.; Hayder, M.F.; El-Sorady, A.E.M. Effect of sowing date of intercropping cowpea with cotton on infestation with some major pests. *Egypt J. Agric. Res.* **1994**, *72*, 691–698.
98. Powell, K.S.; Cooper, P.D.; Forneck, A. The biology, physiology and host-plant interactions of grape phylloxera *Daktulosphaira vitifoliae*. *Behav. Physiol. Root Herbivores* **2013**, *45*, 159–218.
99. Granett, J.; Walker, M.A.; Kocsis, L.; Omer, A.D. Biology and management of grape phylloxera. *Ann. Rev. Entomol.* **2001**, *46*, 387–412. [CrossRef]
100. Rao, N.V.; Rao, C.V.N.; Bhavanil, B.; Naidu, N.V. Influence of intercrops on incidence of early shoot borer, Chiloinfuscatellus Snellen in Sugarcane. *J. Entomol. Res.* **2010**, *34*, 275–276.
101. Bregante, F.; Matta, V.A. Evaluation of the organic system for bean and maize production. *Simiente* **1985**, *55*, 28.
102. Kellow, A.V.; Sedgley, M.; van Heeswijck, R. Interaction between *Vitis vinifera* and grape phylloxera: Changes in root tissue during nodosity formation. *Ann. Bot.* **2004**, *93*, 581–590. [CrossRef]
103. Donald, W.L. Grape Phylloxera [*Daktulosphaira vitifoliae* (Homoptera: Phylloxeridae)] Related Root Damage in Organically and Conventionally Managed Vineyards. Ph.D. Thesis, University of California, Davis, CA, USA, 2000; pp. 12–33.
104. Chitkowski, R.L.; Fisher, J.R. Effect of soil type on the establishment of grape phylloxera colonies in the Pacific Northwest. *Am. J. Enol. Viticult.* **2005**, *56*, 207–211.
105. Wang, Z.Y.; Su, J.P.; Liu, W.W.; Guo, Y.Y. Effects of intercropping vines with tobacco and root extracts of tobacco on grape phylloxera, *Daktulosphaira vitifoliae* Fitch. *J. Integr. Agr.* **2015**, *14*, 1367–1375. [CrossRef]
106. Furlong, M.J.; Wright, D.J.; Dosdall, L.M. Diamondback moth ecology and management: Problems, progress, and prospects. *Ann. Rev. Entomol.* **2013**, *58*, 517–541. [CrossRef] [PubMed]
107. Huang, B.; Shi, Z.H.; Hou, Y.M. Host selection behavior and the fecundity of the diamondback moth, *Plutella xylostella* on multiple host plants. *J. Insect Sci.* **2014**, *14*, 251. [CrossRef] [PubMed]
108. Mucheru-Muna, M.; Pypers, P.; Mugendi, D.; Kung'u, J.; Mugwe, J.; Merck, R.; Vanlauwe, B. A staggered maize-legume intercrop arrangement robustly increases crop yields and economic returns in the highlands of Central Kenya. *Field Crop Res.* **2010**, *115*, 132–139. [CrossRef]
109. Price, P.W. Colonisation of crops by arthropods. Nonequilibrium communities in soybean fields. *Environ. Entomol.* **1976**, *5*, 605–611. [CrossRef]
110. Helenius, J. The influence of mixed intercropping of oats with field beans on the abundance and spatial distribution of cereal aphids (Homoptera: Aphididae). *Agric. Ecos. Environ.* **1989**, *25*, 53–73. [CrossRef]

111. Smith, J.G. Some effects of crop background on populations of aphids and their natural enemies on brussels sprouts. *Ann. Appl. Biol.* **1969**, *63*, 326–330. [CrossRef]
112. Roubinet, E.; Birkhofer, K.; Malsher, G.; Staudacher, K.; Ekbom, B.; Traugott, M.; Jonsson, M. Diet of generalist predators reflects effects of cropping period and farming system on extra- and intraguild prey. *Ecol. Appl.* **2017**, *27*, 1167–1177. [CrossRef]

Article

Grain Yield and Total Protein Content of Organically Grown Oats–Vetch Mixtures Depending on Soil Type and Oats' Cultivar

Katarzyna Pużyńska [1,*], Stanisław Pużyński [2,*], Agnieszka Synowiec [1], Jan Bocianowski [3] and Andrzej Lepiarczyk [1]

[1] Department of Agroecology and Crop Production, Faculty of Agriculture and Economics, University of Agriculture in Krakow, Al. Mickiewicza 21, 31-120 Krakow, Poland; agnieszka.synowiec@urk.edu.pl (A.S.); andrzej.lepiarczyk@urk.edu.pl (A.L.)

[2] Malopolska Agricultural Advisory Centre, ul. Osiedlowa 9, 32-082 Karniowice, Poland

[3] Department of Mathematical and Statistical Methods, Poznań University of Life Sciences, Wojska Polskiego 28, 60-637 Poznań, Poland; jan.bocianowski@up.poznan.pl

* Correspondence: katarzyna.puzynska@urk.edu.pl (K.P.); spuzynski@wp.pl (S.P.); Tel.: +48-12-662-4368 (K.P.)

Abstract: The yield and quality of crop mixtures depend on natural and agrotechnical factors and their relationships. This research aimed to analyze the grain yield, its components and total protein content of the organically grown oat–vetch mixture on two different soils and depending on the oat cultivar. The three-year field experiment with two crop rotations was carried out. The experiment was set up in the southern Poland on two soils: Stagnic Luvisol (S.L.) and Haplic Cambisol (H.C.). One of four oat cultivars ('Celer', 'Furman', 'Grajcar' and 'Kasztan') was grown with the common vetch cv. 'Hanka'. The results showed that the grain yield of mixtures was affected mainly by weather conditions. During the dry season, the share of vetch in the grain yield was 46% lower than in the season of regular rainfall. The share of vetch seeds in the mixture's yield was ca. 21% higher when the mixtures were grown on the S.L. than the H.C. soil. The selection of oats' cultivar for the mixture with vetch affected significantly the thousand seed mass and protein content in the vetch seeds, 46.2–50.4 g and 270–280 g kg^{-1}, respectively. The mixture with Kasztan cultivar yielded the best and this oat cultivar seemed to be the most appropriate for organic conditions; however, in years with high variability of rainfall distribution its usefulness was less.

Keywords: cereal–legume mixture; oats; common vetch; cultivar; soil quality

Citation: Pużyńska, K.; Pużyński, S.; Synowiec, A.; Bocianowski, J.; Lepiarczyk, A. Grain Yield and Total Protein Content of Organically Grown Oats–Vetch Mixtures Depending on Soil Type and Oats' Cultivar. *Agriculture* **2021**, *11*, 79. https://doi.org/10.3390/agriculture11010079

Received: 15 December 2020
Accepted: 15 January 2021
Published: 18 January 2021

Publisher's Note: MDPI stays neutral with regard to jurisdictional claims in published maps and institutional affiliations.

1. Introduction

Cereal–legume mixtures are usually cultivated for grain or green fodder, sometimes as a green manure. Compared to their pure sowing, cereal and legume mixtures are characterized by a higher total protein yield, more stable yielding, especially in unfavorable habitats, a better legume health, and higher nutritional value [1,2]. An additional advantage of the mixture is soil enrichment by legumes with symbiotically fixed nitrogen [3–5]. In the research mixtures of oats with common vetch were tested.

Oat (*Avena sativa* L.) is a cereal with phytosanitary properties in the crop rotation because it is rarely infested by fungal pathogens of stem base and roots [6]. The tolerance of oats to soil acidification, poor soil conditions, low temperature, and higher soil humidity make them a frequent component of many crop rotations, especially in mountainous regions, with a higher share of rainfall [7]. Oats' grain is an excellent feed for horses and dairy cattle because of its chemical composition. Depending on the cultivar, grains of oats contain ca. 100 g kg^{-1} dry matter (d.m.) of total protein, 40–50 g kg^{-1} d.m. of crude fat, 100 g kg^{-1} d.m. of crude fiber, 60 g kg^{-1} d.m. of nitrogen-free extract [8–10]. The biological value of oat protein is not high, but it contains many valuable amino acids, such as lysine and arginine [10]. Of all cereals, oats have the most fiber, mainly in their husks, which reduces their digestibility and energy value [11,12]. Oat products and grain quality can be improved by mixing with legumes [13,14].

Common vetch (*Vicia sativa* L.) contains high amounts of protein in seeds (approx. 33% of dry matter) and vegetative parts, i.e., in straw (approx. 60–120 g kg^{-1} d m.) and green fodder (150–250 g kg^{-1} d.m.) [15]. Vetch seeds can be used as a supplement for animals' feed in the absence or limited access to soybean or cornmeal [16]. Ceglarek et al. [17] underline the high content of thiamine acids and methionine in its protein, in comparison to other legume species. Common vetch is ideal for green forage as it has thin stems rich in fine leaves. The slender shoots of vetch can reach a length of up to 150 cm, so it can easily lodge [18]. Common vetch, like oats, is a good forecrop [19]. However, unlike oats, it has high soil demands. It is also characterized by high water requirements, especially during flowering due to the pile root system and a high transpiration rate [20].

The oat–vetch mixture for grain or green forage production combines the advantages of two different species, e.g., reduced fertilization needs due to symbiotic nitrogen fixation. When mixed with oats, vetch plants are less prone to lodging so that harvesting can be done in one step with a combine harvester. The oat and vetch mixture improves soil structure and growth of succeeding crops. In the mixture, the oat protein complements the vetch's sulfur amino acids, and the vetch protein has a positive effect on the quality of the feed [19,20].

The share of vetch seeds in the mixture with oat is variable [20,21], and for that reason, it is not very popular in cultivation. Moreover, with low rainfall, vetch cannot withstand competition for water with oat, and its share in the mixture yield is small [22]. Another important factor influencing the yield of the mixture are different soil requirements of its components. A proper selection of cultivars for the mixture is essential, especially cereal cultivars characterized by lower competitiveness toward the legume component [21]. To date, there are very few reports in the literature on the effect of cultivar choice on the yield of the cereal and legume mixtures in conditions of organic farming. For this reason, this study aimed to analyze the yield, its components and protein content of grain of four oat cultivars grown organically in a mixture with Hanka's vetch on two different soils.

2. Materials and Methods

2.1. Field Site and Experiment Descriptions

The research was carried out in 2012–2014 in the Experimental Station Mydlniki-Krakow (50°05′ N 19°51′ E) in the southern Poland. The experiment was set up in a randomized block design, with four replications on two types of soils: Stagnic Luvisol (S.L.) and Haplic Cambisol (H.C.) [23], located about 1 km apart. The area of the experiment was under organic farming management since 2009. The description of the soils is given in Table 1. The preceding crop was winter spelt (*Triticum spelta* cv. 'Frankenkorn').

Table 1. Characteristic of the soils.

Parameter	Unit	Stagnic Luvisol	Haplic Cambisol
pH (KCl)	-	6.04	5.31
Total organic C	g kg^{-1}	7.34	6.67
Total N	g kg^{-1}	0.858	0.61
P	mg kg^{-1}	423.0	337.5
K	mg kg^{-1}	148.2	178.3

The mixtures of oat with common vetch (*Vicia sativa*, cv. 'Hanka'; breeder: FN Granum, Wodzierady, Poland) were cultivated for grain. The common vetch was mixed with one of the four oats' cultivars, namely 'Celer', 'Grajcar', 'Kasztan', or 'Furman'. A characteristic of the oats' cultivars is presented in Table 2. The mixtures were sown on 23 March 2012; 16 April 2013; and 20 March 2014, on plots of 18 m^2 (3 × 6 m) area, using plot drill (Hege 80) at row space 13.0 cm. A total of 32 plots were established each year. The planned density of crops was 500 plants m^{-2} of oat and 75 plants m^{-2} of vetch. Crops were cultivated organically.

Table 2. Characteristics of oats' cultivars.

Features	Oats Cultivar			
	'Celer'	**'Grajcar'**	**'Kasztan'**	**'Furman'**
Grain color	yellow	yellow	yellow	yellow
Grain yield	good	good	medium	quite good
Husk share in grain	28.8% (high)	29.5% (very high)	29.4% (very high)	29.0% (high)
Tolerance to soil acidification	average	average	average	quite small
Lodging resistance	average	average	average	big
Recommended sowing rate of seeds (seeds m^{-2})	550–600	550–600	500	400–450
Plant high	quite small	quite small	quite small	medium
No. of days to ripening (since January 1)	198	199	201	206
Thousand grains weight (g)	40.1	35.3	36.9	37.3
Protein content	medium	medium	medium	small to very small
Fat content	medium	medium	very big	small to very small
Areas intended for cultivation	mountainous	mountainous	lowland and mountainous	lowland
Breeder	Małopolska Hodowla Roślin, Sp. z o. o., Poland	Małopolska Hodowla Roślin, Sp. z o. o., Poland	Małopolska Hodowla Roślin, Sp. z o. o., Poland	Hodowla Roślin, Danko, Sp. z o. o., Poland

'Hanka' is a common vetch cultivar of a traditional type of growth, i.e., indeterminate. Plants are lush, rich in leaves ending with sticking tendrils; seeds are brown—thousand seeds weight is 52 g. The cultivar is very fertile, of high total protein content (320 g kg^{-1} d.m.). Tolerance to soil acidification is quite small. It can be grown for seeds, green fodder, or green manure. The cultivar is appropriate for mixing with cereals. Breeder: Firma Nasienna (F.N.) Granum, Poland.

2.2. Measurements

In the early phase of oat growth in BBCH-scale 11–12 (*german* "Biologische Bundesanstalt, Bundessortenamt und CHemische Industrie"), the number of plants per 1 m^2 area was counted to assess mixtures density. Before harvesting, 20 plants were taken for detailed measurements, i.e., the number and weight of panicles, the number of grains, and the 1000 grains weight. Combine harvesting was performed with a plot harvester when oats were fully ripe (BBCH 97–99). After harvesting, grains, and straw of mixtures from the area 18 m^2, were weighed. The final yields of grains per plot were converted into a notional humidity of 15%. For that reason, samples of grains (ca. 40 g.) and straw (ca. 40 g.) were dried at 105 °C using a forced-air oven until a constant weight was obtained. Based on the dry mass values, the grain yields were calculated [24]. Protein content (%) was determined using the InfraXact™ analyzer (Foss, Hillerod, Denmark) based on the near-infrared spectroscopy. The analysis was conducted in three technical replications per sample in the 570–1850 nm wavelengths. Each sample was scanned six times and compared with two internal standards (references) before calculating the mean value.

2.3. Statistical Analysis of the Results

The normality of distribution of the observed traits was tested with Shapiro–Wilk's normality test [25]. Next, the effects of the main factors under study (I factor–soil type: S.L. and H.C.; II factor–oat cultivars: 'Celer', 'Grajcar', 'Kasztan', 'Furman'; III factor–years: 2012, 2013, 2014) as well as all the interactions between them were estimated with a linear model for the three-way analysis of variance (ANOVA) for particular traits. The relationships between the traits were assessed based on Pearson's correlation coefficients and tested with the Tukey's test at $p \leq 0.05$. The results were also analyzed with multivariate methods. The canonical variate analysis (CVA) was applied to present a multi-trait assessment of similarity of the investigated treatments in a lower number of dimensions with the least possible loss of information [26]. This enabled graphic illustration of the variation in the traits of all treatments under analysis. The Mahalanobis distance was suggested to measure multi-trait treatments' similarity [27], whose significance was verified employing critical value *Dcr* known as the least significant distance [28]. Pearson's simple correlation coefficients were estimated to determine each original trait's relative share in the treatments' multivariate variation between values of the first two canonical variates and original individual traits. The GenStat v. 18 statistical software package was used for all the analyses.

The variation coefficient (V) was calculated to characterize the diversity of the sum of rainfall and temperature in the particular months of the growing season (April–August) 2012–2014.

$$V = \frac{S}{\overline{X}} \times 100\%$$

(1)

where:

V—the coefficient of variation,
S—a standard deviation,
$\overline{X}$—arithmetic mean of the variable value.

2.4. Weather Conditions

The weather data were collected from the meteorological station in the Experimental Station in Mydlniki-Kraków (50°05′ N 19°51′ E). The weather conditions during the study period varied (Figures 1–3). The sums of precipitation (Figures 1 and 2) and the average daily air temperature (Figure 3) in 2012–2014 differed from the average for the long-term period (1951–2000). According to [29], the required amount of precipitation for oats during the vegetation period ranges from 270 mm on light (sandy) soils to 400 mm on heavy soils. The water demand for oats increases as the plant develops, reaching the highest values in June and then July. The critical period for water demands for oat in our study was in May 2012, which was very dry, according to the [30] classification. During that month, the amount of rainfall was only 23% of the long-term period. July 2012 was, according to the classification, average—76% of the long-term period and August 2012 was dry—67% of the long-term period. The total rainfall in these months was below the water demand of oat [29]. Based on the humidity characteristics in 2013, April, July, and August were very dry, May humid, and June too humid (213.1 mm of rainfall). In 2014, three out of five months of vegetation were classified as average (April, July and August), May as wet, and June as very dry (43.4 mm of rainfall).

Common vetch also has a high-water demand, especially during the flowering period. In the study period, the temperatures from sowing to harvest were higher than the average for the multi-year period 1951–2000, except for June 2014, when the average temperature was lower by 0.7 °C from the multi-year period. Based on the air temperature classification for Kraków [31], the months of January, March, April, and June 2012 were classified as warm. May, July, and August 2012 were hot. In 2013, January, February, April, and August were classified as regular. March 2013 was very cold, May and June were warm, and July was extremely warm. In 2014, May, June, and August were classified as regular months. April 2014 was warm, and March and July 2014 were extremely warm.

The variation coefficient (V) of the sum of precipitation in individual months of the vegetation period in 2012 was equal to 26%, proving the average variability of rainfall in that period. In 2013, the V was equivalent to 107%, which shows a substantial variability. In 2014, the V in individual months was 41%, which denotes a large variability of precipitation. Temperature variability in the respective months of vegetation period 2012–2014 was different. The V of temperature for the growing season 2012 was 70%, which denotes a large variability. In 2013, V = 28%, and in 2014, 25% indicated the average variability of temperature.

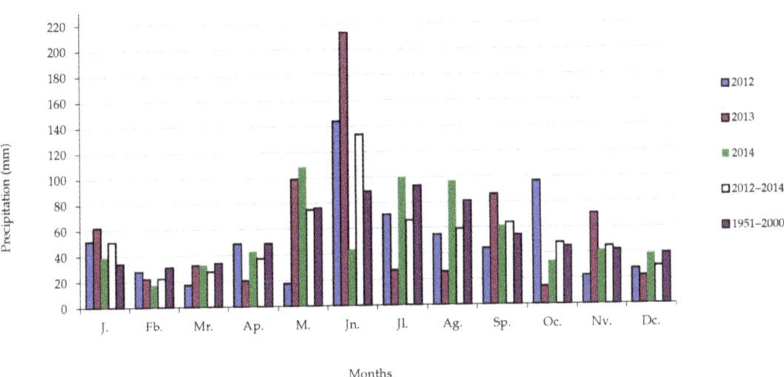

Figure 1. Sum of precipitation (mm) in particular months of 2012–2014 and multiyear 1951–2000.

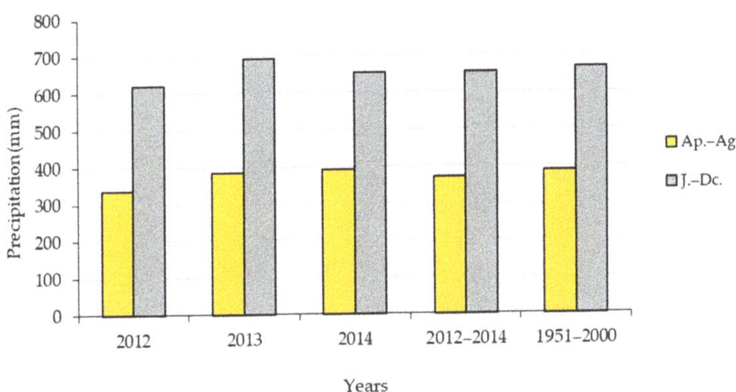

Figure 2. Sum of precipitation (mm) in the vegetative period (April–August) and the years of study 2012–2014 compared to multiyear.

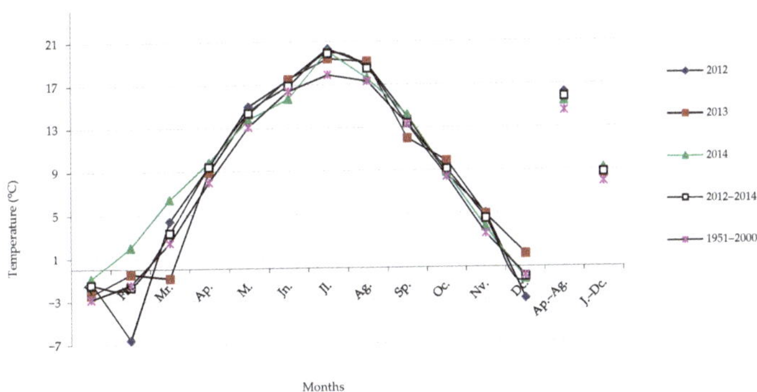

Figure 3. Mean temperatures (°C) in the months of 2012–2014 and in multiyear 1951–2000.

3. Results

In our study, all 13 quantitative traits had a normal distribution. The ANOVA indicated a statistically significant influence of soil type, years, cultivars, and the year × cultivar and year × soil type interactions for all 13 traits (Table 3). The soil type and soil type × cultivar interactions were not significant only for the tiller number. The year × soil type × cultivar was significant for all traits except panicle number (Table 3).

Table 3. Mean squares from three-way analysis of variance for observed traits.

Source of Variation	d.f.	o. Protein	v. Protein	Grain No.	Yield	v. Share	Panicle No.	Tiller no.	TWG	TSW	Panicle g.w.	o. Plant No.	v. Plant No.	o. Height
Replication	3	0.046	0.1701	1.693	0.031	18.1	12,029	0.0003	2.135	2.308	0.0007	22.26	10.19	4.83
ST	1	598.8 ***	148.9 ***	567.3 ***	35.02 ***	10,372 ***	198,586 **	0.0113	9.388 *	972.8 **	0.534 ***	168.01 ***	177.85 *	2889 ***
Residual 1	3	0.044	0.031	0.366	0.058	42.45	3802	0.0023	0.652	6.638	0.003	0.2	17.05	3.258
Cultivar	3	741.4 ***	421.4 ***	65.76 ***	0.281 ***	234.43 ***	77,537 ***	0.115 ***	203.9 ***	83.8 ***	0.048 ***	4623.3 ***	166.4 ***	135.2 ***
ST × Cultivar	3	160.5 ***	132.4 ***	92.08 ***	0.607 ***	189.93 ***	112,019 ***	0.008	7.271 **	30.5 **	0.071 ***	2717.5 ***	32.3 *	117.9 ***
Residual 2	18	0.913	0.59	2.258	0.0238	14.58	3267	0.003	1.002	5.82	0.004	36.57	8.82	4.35
Year	2	20,446.5 ***	446.5 ***	134.6 ***	13.43 ***	19,292 ***	2,113,463 ***	0.140 ***	137.1 ***	931.4 ***	0.199 ***	131,550 ***	682.6 ***	335.4 ***
Year x S.T.	2	20.79 ***	883.3 ***	87.60 ***	1.057 ***	342.1 ***	41,734 ***	0.131 ***	42.69 ***	344.2 ***	0.064 ***	10,784 ***	1438.1 ***	183.8 ***
Year × Cultivar	6	308.7 ***	92.9 ***	46.5 ***	0.219 ***	272.67 ***	58,217 ***	0.075 ***	4.169 ***	25.56 *	0.022 ***	45,497 ***	287.3 ***	44.13 ***
Year × ST × Cultivar	6	526.9 ***	255.6 ***	29.87 ***	0.322 ***	208.25 ***	8441	0.038 ***	12.59 ***	52.41 ***	0.025 ***	5523 ***	126.7 ***	42.55 ***
Residual 3	48	1.02	0.468	1.79	0.0336	18.21	4382	0.0029	0.53	8.788	0.003	24.44	7.338	6.19

Abbreviations: ST—soil type; d.f.—degrees of freedom; o. Protein—protein content in oats grain (g·kg^{-1}); v. Protein—protein content in vetch seeds (g·kg^{-1}); Grain No.—number of oats grains (pcs.) per panicle; Yield—mixtures yield; v. Share—the share of vetch in the mixture's yield (%); Panicle No.—number of oats panicles per m^2; Tiller no.—number of oats' tillers; TWG—thousand grain mass of oats; TSW—thousand grain mass of vetch; Panicle g.w.—mass grains (g) per oats panicle; o. Plant No.—density of oats after spring emergence (pcs m^2); v. Plant No.—density of common vetch (pcs m^2) spring; o. Height—the height of oats canopy (cm). * $p < 0.05$; ** $p < 0.01$; *** $p < 0.001$; d.f.–the number of degrees of freedom.

3.1. Selected Biometric Features of the Mixture

The average spring density of oat was similar for both soil types and the majority of oats' cultivars (Table 4). The cultivar factor as well as weather during emergency of the oat significantly affected its density.

Table 4. Oats' density in spring (pieces m^{-2}) in mixture depending on the soil type (factor I), oat cultivar (factor II), and study years (factor III).

Soil Type	Years	Oat Cultivar				Mean [1] ± SD [3]
		'Celer'	'Furman'	'Grajcar'	'Kasztan'	
Stagnic Luvisol	2012	500	460	483	497	485 ± 18.5
	2013	364	493	460	474	448 ± 57.8
	2014	409	405	452	447	428 ± 25.0
	Mean [2] ± SD [3]	424 ± 69.6	452 ± 44.8	465 ± 16.0	473 ± 25.0	454 ns
Haplic Cambisol	2012	490	439	441	488	464 ± 28.4
	2013	452	500	498	485	484 ± 22.2
	2014	417	443	399	390	412 ± 23.4
	Mean [2] ± SD [3]	453 ± 36.3	461 ± 34.3	446 ± 49.7	454 ± 55.6	453 ns
Mean	2012	495	449	462	493	475 ± 22.7 x
	2013	408	497	479	479	466 ± 39.5 y
	2014	413	424	426	419	420 ± 5.8 z
	Mean [2] ± SD [3]	438 ± 48.9 b	456 ± 37.0 a	456 ± 27.3 a	463 ± 39.5 a	453
LSD $_{0.05}$ soil type		ns [4]				
LSD $_{0.05}$ cultivar		8.03				
LSD $_{0.05}$ years		5.14				
LSD $_{0.05}$ soil type × cultivar		11.2				
LSD $_{0.05}$ soil type × years		7.27				
LSD $_{0.05}$ cultivar × years		10.3				

[1] Mean for the soil type, regardless of the oat cultivar; [2] Mean for the year 2012–2014; [3] S.D.—standard deviation; [4] ns—non-significant. Homogeneous groups were created for the main factors. According to Tukey's test, mean values marked with the same letters do not differ significantly ($p \leq 0.05$). Small letters (a, b, c) for mean values of the second-factor levels—oats cultivars and x, y and z letters for the third-factor levels—study years were chosen. The three-factor ANOVA—first-factor, soil type: Stagnic Luvisol and Haplic Cambisol; second-factor, oat cultivar: 'Celer', 'Furman', 'Grajcar', 'Kasztan'; third-factor, years: 2012, 2013, 2014.

The number of oat tillers in the mixtures was low and similar, regardless of the soil types (Table 5). However, the oat cultivars in the mixtures tilled differently, with cv. 'Celer', which developed the highest number of tillers, especially in 2013, and cv. 'Kasztan'—the lowest (1.14). The lowest oats' tillering was noted in 2014; it was 10% lower than in 2013.

Table 5. Number of oats' tillers in mixtures depending on the soil type (factor I), oat cultivar (factor II), and study years (factor III).

Soil Type	Years	Oat Cultivar				Mean [1] ± SD [3]
		'Celer'	'Furman'	'Grajcar'	'Kasztan'	
Stagnic Luvisol	2012	1.22	1.11	1.21	1.09	1.16 ± 0.07
	2013	1.68	1.34	1.17	1.16	1.34 ± 0.24
	2014	1.03	1.20	1.10	1.08	1.10 ± 0.07
	Mean [2] ± SD [3]	1.31 ± 0.33	1.22 ± 0.11	1.16 ± 0.06	1.11 ± 0.05	1.20 ns
Haplic Cambisol	2012	1.28	1.33	1.32	1.15	1.27 ± 0.08
	2013	1.34	1.21	1.12	1.18	1.21 ± 0.09
	2014	1.25	1.18	1.10	1.20	1.18 ± 0.06
	Mean [2] ± SD [3]	1.29 ± 0.05	1.24 ± 0.08	1.18 ± 0.12	1.18 ± 0.02	1.22 ns
Mean	2012	1.25	1.22	1.26	1.12	1.21 ± 0.06 y
	2013	1.51	1.28	1.14	1.17	1.27 ± 0.17 x
	2014	1.14	1.19	1.10	1.14	1.14 ± 0.04 z
	Mean [2] ± SD [3]	1.30 ± 0.19 a	1.23 ± 0.04 b	1.17 ± 0.09 c	1.14 ± 0.02 c	1.21

Table 5. *Cont.*

Soil Type	Years	Oat Cultivar				Mean [1] ± SD [3]
		'Celer'	'Furman'	'Grajcar'	'Kasztan'	
LSD $_{0.05}$ soil type				ns [4]		
LSD $_{0.05}$ cultivar				0.045		
LSD $_{0.05}$ years				0.033		
LSD $_{0.05}$ soil type × cultivar				ns		
LSD $_{0.05}$ soil type × years				0.046		
LSD $_{0.05}$ cultivar × years				0.065		

[1] Mean for the soil type, regardless of the oat cultivar; [2] Mean for the year 2012–2014; [3] S.D.—standard deviation; [4] ns—non-significant. Homogeneous groups were created for the main factors. According to Tukey's test, mean values marked with the same letters do not differ significantly ($p \leq 0.05$). Small letters (a, b, c) for mean values of the second-factor levels—oats cultivars and x, y and z letters for the third-factor levels—study years were chosen. The three-factor ANOVA—first-factor, soil type: Stagnic Luvisol and Haplic Cambisol; second-factor, oat cultivar: 'Celer', 'Furman', 'Grajcar', 'Kasztan'; third-factor, years: 2012, 2013, 2014.

Vetch density in the mixtures, as counted in spring, was ca. 30% lower than the planned one (Table 6). A higher density was noted on the H.C. soil than the S.L. soil. The vetch density depended on selected oats cultivar for the mixture and varied between 49.1 for cv. 'Furman' to 55.3 pieces m^{-2} for cv. 'Celer'. The highest vetch densities in the mixtures were found in 2013 year whereas the lowest in 2014.

Table 6. Vetch density in spring (pieces m^{-2}) in mixture depending on the soil type (factor I), oat cultivar (factor II), and study years (factor III).

Soil Type	Years	Oat Cultivar				Mean [1] ± SD [3]
		'Celer'	'Furman'	'Grajcar'	'Kasztan'	
Stagnic Luvisol	2012	52.0	64.0	56.5	52.0	56.1 ± 5.66
	2013	55.0	42.5	40.0	51.0	47.1 ± 7.05
	2014	51.0	41.0	61.0	49.3	50.6 ± 8.21
	Mean [2] ± SD [3]	52.7 ± 2.08	49.2 ± 12.9	52.5 ± 11.1	50.8 ± 1.37	51.3 B
Haplic Cambisol	2012	58.0	51.0	55.0	46.0	52.5 ± 5.20
	2013	69.0	64.0	63.0	65.0	65.3 ± 2.63
	2014	47.0	32.0	46.0	52.0	44.3 ± 8.58
	Mean [2] ± SD [3]	58.0 ± 11.0	49.0 ± 16.1	54.7 ± 8.50	54.3 ± 9.71	54.0 A
Mean	2012	55.0	57.5	55.8	49.0	54.3 ± 3.70 y
	2013	62.0	53.3	51.5	58.0	56.2 ± 4.74 x
	2014	49.0	36.5	53.5	50.7	47.4 ± 7.52 z
	Mean [2] ± SD [3]	55.3 ± 6.51 a	49.1 ± 11.1 b	53.6 ± 2.15 ab	52.6 ± 4.78 ab	52.6
LSD $_{0.05}$ soil type				2.68		
LSD $_{0.05}$ cultivar				2.42		
LSD $_{0.05}$ years				1.68		
LSD $_{0.05}$ soil type × cultivar				3.43		
LSD $_{0.05}$ soil type × years				2.32		
LSD $_{0.05}$ cultivar × years				3.27		

[1] Mean for the soil type, regardless of the oat cultivar; [2] Mean for the year 2012–2014; [3] S.D.—standard deviation; Homogeneous groups were created for the main factors. According to Tukey's test, mean values marked with the same letters do not differ significantly ($p \leq 0.05$). Capital letters (A and B) for mean values of the first factor levels—soil types, small letters (a, b, c) for mean values of the second-factor levels—oats cultivars and x, y and z letters for the third-factor levels—study years were chosen. The three-factor ANOVA—first-factor, soil type: Stagnic Luvisol and Haplic Cambisol; second-factor, oat cultivar: 'Celer', 'Furman', 'Grajcar', 'Kasztan'; third-factor, years: 2012, 2013, 2014.

3.2. Yield of Mixtures

On average, the mixture yielded 40% lower on Haplic Cambisol (H.C.), compared to Stagnic Luvisol (S.L.) (Table 7). The yield of three oat cultivars' grown with common vetch on the S.L. soil was 3.06—3.19 t ha^{-1}, except for cv. 'Grajcar' that yielded significantly lower. On the H.C. soil, the yield of cv. 'Kasztan' was by 0.2—0.46 t ha^{-1} higher compared to other cultivars. The yielding of oat cultivars with vetch varied between years. The highest yields of the mixtures were in dry 2012, and the lowest, in regular 2014.

Table 7. Seed yield (t ha^{-1}) of the mixture depending on the soil type (factor I), oat cultivar (factor II), and study years (factor III).

Soil Type	Years	Oat Cultivar				Mean [1] $\pm$ SD [3]
		'Celer'	'Furman'	'Grajcar'	'Kasztan'	
Stagnic Luvisol	2012	4.15	3.84	3.61	3.79	3.84 $\pm$ 0.23
	2013	2.53	2.43	2.39	2.22	2.39 $\pm$ 0.13
	2014	2.69	3.30	2.31	3.17	2.87 $\pm$ 0.46
	Mean [2] $\pm$ SD [3]	3.12 $\pm$ 0.89	3.19 $\pm$ 0.71	2.77 $\pm$ 0.73	3.06 $\pm$ 0.79	3.03 A
Haplic Cambisol	2012	1.78	2.09	2.47	2.81	2.29 $\pm$ 0.45
	2013	1.07	1.12	1.18	1.24	1.15 $\pm$ 0.07
	2014	2.00	1.99	1.99	2.18	2.04 $\pm$ 0.09
	Mean [2] $\pm$ SD [3]	1.62 $\pm$ 0.49	1.73 $\pm$ 0.53	1.88 $\pm$ 0.65	2.08 $\pm$ 0.79	1.83 B
Mean	2012	2.96	2.96	3.04	3.30	3.07 $\pm$ 0.16 x
	2013	1.80	1.78	1.78	1.73	1.77 $\pm$ 0.03 z
	2014	2.35	2.64	2.15	2.68	2.45 $\pm$ 0.25 y
	Mean [2] $\pm$ SD [3]	2.37 $\pm$ 0.58 bc	2.46 $\pm$ 0.61 ab	2.32 $\pm$ 0.65 c	2.57 $\pm$ 0.79 a	2.43
LSD $_{0.05}$ soil type				0.157		
LSD $_{0.05}$ cultivar				0.126		
LSD $_{0.05}$ years				0.111		
LSD $_{0.05}$ soil type $\times$ cultivar				0.178		
LSD $_{0.05}$ soil type $\times$ years				0.157		
LSD $_{0.05}$ cultivar $\times$ years				0.220		

[1] Mean for the soil type, regardless of the oat cultivar; [2] Mean for the year 2012–2014; [3] S.D.—standard deviation. Homogeneous groups were created for the main factors. According to Tukey's test, mean values marked with the same letters do not differ significantly ($p \leq 0.05$). Capital letters (A and B) for mean values of the first factor levels—soil types, small letters (a, b, c) for mean values of the second-factor levels—oats cultivars and x, y and z letters for the third-factor levels—study years were chosen. The three-factor ANOVA—first-factor, soil type: Stagnic Luvisol and Haplic Cambisol; second-factor, oat cultivar: 'Celer', 'Furman', 'Grajcar', 'Kasztan'; third-factor, years: 2012, 2013, 2014.

The share of vetch seeds in the mixtures was variable. On average, it was 20% higher on the H.C. soil than the S.L. soil (Table 8). It was also highest in 2013 (65.7%) and the lowest–in a dry 2012 (19.6%). The vetch seed's share also depended on the selected oats cultivar and was the highest for cv. 'Grajcar', and the lowest for cv. 'Kasztan'.

Table 8. Share (%) of common vetch seeds in the mixture depending on the soil type (factor I), oat cultivar (factor II), and study years (factor III).

Soil Type	Years	Oat Cultivar				Mean [1] $\pm$ SD [3]
		'Celer'	'Furman'	'Grajcar'	'Kasztan'	
Stagnic Luvisol	2012	22.3	35.1	29.7	20.8	27.0 $\pm$ 6.67
	2013	75.4	75.0	76.4	76.0	75.7 $\pm$ 0.65
	2014	78.1	69.7	65.9	70.7	71.1 $\pm$ 5.13
	Mean [2] $\pm$ SD [3]	58.6 $\pm$ 31.5	59.9 $\pm$ 21.7	57.3 $\pm$ 24.5	55.8 $\pm$ 30.5	57.9 A
Haplic Cambisol	2012	4.8	23.0	12.7	8.5	12.2 $\pm$ 7.86
	2013	61.6	55.1	61.7	44.7	55.8 $\pm$ 8.00
	2014	41.9	31.0	60.2	40.4	43.4 $\pm$ 12.2
	Mean [2] $\pm$ SD [3]	36.1 $\pm$ 28.8	36.4 $\pm$ 16.7	44.9 $\pm$ 27.9	31.2 $\pm$ 19.8	37.1 B
Mean	2012	13.6	29.1	21.2	14.6	19.6 $\pm$ 7.14 z
	2013	68.5	65.0	69.1	60.4	65.7 $\pm$ 4.00 x
	2014	60.0	50.4	63.0	55.6	57.2 $\pm$ 5.52 y
	Mean [2] $\pm$ SD [3]	47.4 $\pm$ 29.6 b	48.2 $\pm$ 18.1 ab	51.1 $\pm$ 26.1 a	43.5 $\pm$ 25.1 c	47.5

Table 8. *Cont.*

Soil Type	Years	Oat Cultivar				Mean [1] $\pm$ SD [3]
		'Celer'	'Furman'	'Grajcar'	'Kasztan'	
LSD $_{0.05}$ soil type				4.23		
LSD $_{0.05}$ cultivar				3.12		
LSD $_{0.05}$ years				2.58		
LSD $_{0.05}$ soil type $\times$ cultivar				4.41		
LSD $_{0.05}$ soil type $\times$ years				3.65		
LSD $_{0.05}$ cultivar $\times$ years				5.16		

[1] Mean for the soil type, regardless of the oat cultivar; [2] Mean for the year 2012–2014; [3] S.D.—standard deviation. Homogeneous groups were created for the main factors. According to Tukey's test, mean values marked with the same letters do not differ significantly ($p \leq 0.05$). Capital letters (A and B) for mean values of the first factor levels—soil types, small letters (a, b, c) for mean values of the second-factor levels—oats cultivars and x, y and z letters for the third-factor levels—study years were chosen. The three-factor ANOVA—first-factor, soil type: Stagnic Luvisol and Haplic Cambisol; second-factor, oat cultivar: 'Celer', 'Furman', 'Grajcar', 'Kasztan'; third-factor, years: 2012, 2013, 2014.

The oat–vetch mixture's straw yield was significantly differentiated by the examined factors and their interaction (Table 9). A substantially higher straw yield was found on the S.L. soil (4.72 t ha^{-1}) than the H.C. soil (3.72 t ha^{-1}). Contrary to the grains' yield, the highest straw yield was recorded in 2014 (5.58 t ha^{-1}), and the lowest in 2012 (3.13 t ha^{-1}).

Table 9. Straw yield (t ha^{-1}) for oats-vetch mixtures depending on the soil type (factor I), oat cultivar (factor II), and study years (factor III).

Soil Type	Years	Oat Cultivar				Mean [1] $\pm$ SD [3]
		'Celer'	'Furman'	'Grajcar'	'Kasztan'	
Stagnic Luvisol	2012	3.99	3.80	3.79	3.30	3.72 $\pm$ 0.29
	2013	5.82	5.14	4.71	4.59	5.06 $\pm$ 0.56
	2014	5.47	5.99	4.59	5.53	5.39 $\pm$ 0.59
	Mean [2] $\pm$ SD [3]	5.09 $\pm$ 0.97	4.98 $\pm$ 1.11	4.36 $\pm$ 0.50	4.47 $\pm$ 1.12	4.72 A
Haplic Cambisol	2012	3.00	2.73	2.27	2.17	2.54 $\pm$ 0.39
	2013	2.69	3.10	2.82	2.84	2.86 $\pm$ 0.17
	2014	5.73	6.48	5.62	5.20	5.76 $\pm$ 0.53
	Mean [2] $\pm$ SD [3]	3.81 $\pm$ 1.67	4.10 $\pm$ 2.07	3.57 $\pm$ 1.80	3.41 $\pm$ 1.59	3.72 B
Mean	2012	3.49	3.26	3.03	2.74	3.13 $\pm$ 0.32 z
	2013	4.26	4.12	3.76	3.71	3.96 $\pm$ 0.27 y
	2014	5.60	6.24	5.10	5.36	5.58 $\pm$ 0.49 x
	Mean [2] $\pm$ SD [3]	4.45 $\pm$ 1.07 a	4.54 $\pm$ 1.53 a	3.96 $\pm$ 1.05 b	3.94 $\pm$ 1.33 b	4.22
LSD $_{0.05}$ soil type				0.166		
LSD $_{0.05}$ cultivar				0.278		
LSD $_{0.05}$ years				0.240		
LSD $_{0.05}$ soil type $\times$ cultivar				ns [4]		
LSD $_{0.05}$ soil type $\times$ years				0.321		
LSD $_{0.05}$ cultivar $\times$ years				0.479		

[1] Mean for the soil type, regardless of the oat cultivar; [2] Mean for the year 2012–2014; [3] S.D.—standard deviation; [4] ns—non-significant. Homogeneous groups were created for the main factors. According to Tukey's test, mean values marked with the same letters do not differ significantly ($p \leq 0.05$). Capital letters (A and B) for mean values of the first factor levels—soil types, small letters (a, b, c) for mean values of the second-factor levels—oats cultivars and x, y and z letters for the third-factor levels—study years were chosen. The three-factor ANOVA—first-factor, soil type: Stagnic Luvisol and Haplic Cambisol; second-factor, oat cultivar: 'Celer', 'Furman', 'Grajcar', 'Kasztan'; third-factor, years: 2012, 2013, 2014.

3.3. Selected Components of Yield Structure

A substantially greater number of oats' panicles was found on H.C. soil (330 pieces m^{-2}) than the S.L. soil (285 pieces m^{-2}) (Table 10). On average, in the mixtures, the largest number of panicles developed cv. 'Celer' (344 pieces m^{-2}) and the smallest—cv. 'Grajcar' (282 pieces m^{-2}). Interestingly, during the dry 2012 year, oat developed almost twice more panicles than in the regular year 2014. In that year, regardless of the soil type, cv. 'Celer' developed the highest number of panicles (559—535 pieces m^{-2}). The number of oat panicles per m^{-2} decreased in the following years, most probably resulting from a continuous sequence of cereals in the crop rotation, and lack of fertilization.

Table 10. Number of oat panicles (pieces m^{-2}) in the mixtures depending on the soil type (factor I), oat cultivar (factor II), and study years (factor III).

Soil Type	Years	Oat Cultivar				Mean [1] $\pm$ SD [3]
		'Celer'	'Furman'	'Grajcar'	'Kasztan'	
Stagnic Luvisol	2012	511	388	455	453	452 $\pm$ 50.3
	2013	203	262	219	180	216 $\pm$ 34.6
	2014	151	227	201	170	187 $\pm$ 33.7
	Mean [2] $\pm$ SD [3]	288 $\pm$ 194.6	292 $\pm$ 84.7	292 $\pm$ 141.8	268 $\pm$ 160.6	285 B
Haplic Cambisol	2012	559	350	415	516	460 $\pm$ 95.0
	2013	316	223	225	291	264 $\pm$ 47.0
	2014	325	283	176	286	268 $\pm$ 63.9
	Mean [2] $\pm$ SD [3]	400 $\pm$ 137.7	285 $\pm$ 63.5	272 $\pm$ 126.2	364 $\pm$ 131.3	330 A
Mean	2012	535	369	435	485	456 $\pm$ 70.9 x
	2013	260	243	222	236	240 $\pm$ 15.7 y
	2014	238	255	189	228	227 $\pm$ 28.2 y
	Mean [2] $\pm$ SD [3]	344 $\pm$ 165.6 a	289 $\pm$ 69.7 c	282 $\pm$ 133.7 c	316 $\pm$ 146.0 b	308
LSD $_{0.05}$ soil type		20.0				
LSD $_{0.05}$ cultivar		23.3				
LSD $_{0.05}$ years		20.0				
LSD $_{0.05}$ soil type $\times$ cultivar		32.9				
LSD $_{0.05}$ soil type $\times$ years		28.3				
LSD $_{0.05}$ cultivar $\times$ years		40.0				

[1] Mean for the soil type, regardless of the oat cultivar; [2] Mean for the year 2012–2014; [3] S.D.—standard deviation. Homogeneous groups were created for the main factors. According to Tukey's test, mean values marked with the same letters do not differ significantly ($p \leq 0.05$). Capital letters (A and B) for mean values of the first factor levels—soil types, small letters (a, b, c) for mean values of the second-factor levels—oats cultivars and x, y and z letters for the third-factor levels—study years were chosen. The three-factor ANOVA—first-factor, soil type: Stagnic Luvisol and Haplic Cambisol; second-factor, oat cultivar: 'Celer', 'Furman', 'Grajcar', 'Kasztan'; third-factor, years: 2012, 2013, 2014.

Significantly more oats' grains per panicle (GPP), by 32%, were found on the S.L. soil, compared to the H.C. soil (Table 11). The number of GPP differed significantly for the oats' cultivars and was in a range of 10.5 for cv. 'Grajcar' to 14.0 for cv. 'Kasztan'. Contrary to the number of panicles per m^{-2}, oat developed 13% more GPP in 2014 than in 2012.

Table 11. Number of grains (pieces) per oat panicle in the mixtures depending on the soil type (factor I), oat cultivar (factor II), and study years (factor III).

Soil Type	Years	Oat Cultivar				Mean [1] $\pm$ SD [3]
		'Celer'	'Furman'	'Grajcar'	'Kasztan'	
Stagnic Luvisol	2012	11.9	20.5	13.6	19.7	16.4 $\pm$ 4.31
	2013	17.7	17.8	7.7	10.9	13.5 $\pm$ 5.08
	2014	18.6	16.0	12.3	13.8	15.2 $\pm$ 2.75
	Mean [2] $\pm$ SD [3]	16.1 $\pm$ 3.63	18.1 $\pm$ 2.28	11.2 $\pm$ 3.10	14.8 $\pm$ 4.46	15.0 A
Haplic Cambisol	2012	4.6	8.1	12.2	11.1	9.0 $\pm$ 3.43
	2013	6.7	6.4	7.0	9.9	7.5 $\pm$ 1.60
	2014	12.6	14.7	10.5	18.4	14.0 $\pm$ 3.37
	Mean [2] $\pm$ SD [3]	8.0 $\pm$ 4.15	9.7 $\pm$ 4.40	9.9 $\pm$ 2.67	13.1 $\pm$ 4.60	10.2 B
Mean	2012	8.2	14.3	12.9	15.4	12.7 $\pm$ 3.15 y
	2013	12.2	12.1	7.3	10.4	10.5 $\pm$ 2.28 z
	2014	15.6	15.4	11.4	16.1	14.6 $\pm$ 2.18 x
	Mean [2] $\pm$ SD [3]	12.0 $\pm$3.68 b	13.9 $\pm$1.65 a	10.5 $\pm$2.88 c	14.0 $\pm$3.12 a	12.6

Table 11. *Cont.*

Soil Type	Years	Oat Cultivar				Mean [1] ± SD [3]
		'Celer'	'Furman'	'Grajcar'	'Kasztan'	
LSD $_{0.05}$ soil type				0.393		
LSD $_{0.05}$ cultivar				1.28		
LSD $_{0.05}$ years				0.809		
LSD $_{0.05}$ soil type × cultivar				1.62		
LSD $_{0.05}$ soil type × years				1.01		
LSD $_{0.05}$ cultivar × years				1.62		

[1] Mean for the soil type, regardless of the oat cultivar; [2] Mean for the year 2012–2014; [3] S.D.—standard deviation. Homogeneous groups were created for the main factors. According to Tukey's test, mean values marked with the same letters do not differ significantly ($p \leq 0.05$). Capital letters (A and B) for mean values of the first factor levels—soil types, small letters (a, b, c) for mean values of the second-factor levels—oats cultivars and x, y and z letters for the third-factor levels—study years were chosen. The three-factor ANOVA—first-factor, soil type: Stagnic Luvisol and Haplic Cambisol; second-factor, oat cultivar: 'Celer', 'Furman', 'Grajcar', 'Kasztan'; third-factor, years: 2012, 2013, 2014.

The soil type significantly differentiated the mass of 1000 grains (MTG) of oat in the mixture with vetch (Table 12). The MTGs of the oats' cultivars in this experiment was lower than standard values (Table 2). A greater MTG was found for oats on the H.C. soil than the S.L. soil. The oat cultivars also differed in the MTG, which was in a range of 31.8–38.7 g for mixture with cv. 'Grajcar' and cv. 'Celer', respectively. In 2013, the oat MTG was 16% higher than in 2012.

Table 12. 1000-grain mass (g) of oat in the mixture with vetch depending on the soil type (factor I), oat cultivar (factor II), and study years (factor III).

Soil Type	Years	Oat Cultivar				Mean [1] ± SD [3]
		'Celer'	'Furman'	'Grajcar'	'Kasztan'	
Stagnic Luvisol	2012	36.4	32.4	28.2	33.4	32.6 ± 3.39
	2013	41.2	37.0	36.4	37.8	38.1 ± 2.16
	2014	36.9	33.4	29.7	32.6	33.1 ± 2.96
	Mean [2] ± SD [3]	38.2 ± 2.66	34.2 ± 2.40	31.4 ± 4.38	34.6 ± 2.78	34.6 B
Haplic Cambisol	2012	36.8	30.1	33.1	34.3	33.6 ± 2.75
	2013	40.8	34.7	31.1	38.5	36.3 ± 4.27
	2014	40.0	35.3	32.1	36.1	35.9 ± 3.24
	Mean [2] ± SD [3]	39.2 ± 2.12	33.3 ± 2.80	32.1 ± 1.01	36.3 ± 2.11	35.2 A
Mean	2012	36.6	31.3	30.7	33.9	33.1 ± 2.71 z
	2013	41.0	35.8	33.8	38.1	37.2 ± 3.11 x
	2014	38.4	34.3	30.9	34.3	34.5 ± 3.08 y
	Mean [2] ± SD [3]	38.7 ± 2.21 a	33.8 ± 2.32 c	31.8 ± 1.72 d	35.4 ± 2.34 b	34.9
LSD $_{0.05}$ soil type				0.525		
LSD $_{0.05}$ cultivar				0.816		
LSD $_{0.05}$ years				0.441		
LSD $_{0.05}$ soil type × cultivar				1.13		
LSD $_{0.05}$ soil type × years				0.624		
LSD $_{0.05}$ cultivar × years				0.883		

[1] Mean for the soil type, regardless of the oat cultivar; [2] Mean for the year 2012–2014; [3] S.D.—standard deviation. Homogeneous groups were created for the main factors. According to Tukey's test, mean values marked with the same letters do not differ significantly ($p \leq 0.05$). Capital letters (A and B) for mean values of the first factor levels—soil types, small letters (a, b, c) for mean values of the second-factor levels—oats cultivars and x, y and z letters for the third-factor levels—study years were chosen. The three-factor ANOVA—first-factor, soil type: Stagnic Luvisol and Haplic Cambisol; second-factor, oat cultivar: 'Celer', 'Furman', 'Grajcar', 'Kasztan'; third-factor, years: 2012, 2013, 2014.

The significant relationships of the mass of grains per oats' panicle were similar to the relationships presented for the MTG of oats (Table 13).

Table 13. Mass grains (g) per oat panicle in the oat–vetch mixture depending on the soil type (factor I), oat cultivar (factor II), and study years (factor III).

Soil Type	Years	Oat Cultivar				Mean [1] ± SD [3]
		'Celer'	'Furman'	'Grajcar'	'Kasztan'	
Stagnic Luvisol	2012	0.43	0.66	0.41	0.65	0.54 ± 0.14
	2013	0.43	0.47	0.25	0.28	0.36 ± 0.11
	2014	0.48	0.47	0.42	0.42	0.45 ± 0.03
	Mean [2] ± SD [3]	0.45 ± 0.03	0.53 ± 0.11	0.36 ± 0.09	0.45 ± 0.18	0.45 A
Haplic Cambisol	2012	0.17	0.24	0.40	0.39	0.30 ± 0.12
	2013	0.17	0.18	0.22	0.25	0.21 ± 0.04
	2014	0.31	0.42	0.32	0.50	0.39 ± 0.09
	Mean [2] ± SD [3]	0.22 ± 0.08	0.28 ± 0.12	0.32 ± 0.09	0.38 ± 0.13	0.30 B
Mean	2012	0.30	0.45	0.41	0.52	0.42 ±0.09 x
	2013	0.30	0.33	0.24	0.27	0.28 ±0.04 y
	2014	0.40	0.44	0.37	0.46	0.42 ±0.04 x
	Mean [2] ± SD [3]	0.33 ± 0.06 b	0.41 ± 0.07 a	0.34 ± 0.09 b	0.42 ± 0.13 a	0.37
LSD $_{0.05}$ soil type				0.034		
LSD $_{0.05}$ cultivar				0.048		
LSD $_{0.05}$ years				0.035		
LSD $_{0.05}$ soil type × cultivar				0.068		
LSD $_{0.05}$ soil type × years				0.050		
LSD $_{0.05}$ cultivar × years				0.070		

[1] Mean for the soil type, regardless of the oat cultivar; [2] Mean for the year 2012–2014; [3] S.D.—standard deviation. Homogeneous groups were created for the main factors. According to Tukey's test, mean values marked with the same letters do not differ significantly ($p \leq 0.05$). Capital letters (A and B) for mean values of the first factor levels—soil types, small letters (a, b, c) for mean values of the second-factor levels—oats cultivars and x, y and z letters for the third-factor levels—study years were chosen. The three-factor ANOVA—first-factor, soil type: Stagnic Luvisol and Haplic Cambisol; second-factor, oat cultivar: 'Celer', 'Furman', 'Grajcar', 'Kasztan'; third-factor, years: 2012, 2013, 2014.

Relative to oats, the mass of 1000 seeds (MTS) of vetch was 12% higher on the S.L. soil than the H.C. soil (Table 14). The MTS of vetch was also considerably influenced by the cultivar of oat, as the mixture companion. The highest MTS of vetch was found in the mixture with oat cv. 'Grajcar', and the lowest in the mixture with oat cv. 'Kasztan'. Moreover, the MTS of vetch varied significantly over the years of the study. The highest MTS of vetch was in a regular 2014, and the lowest in a dry 2012.

Table 14. 1000-seed mass (g) of vetch cv. 'Hanka' of the mixture depending on the soil type (factor I), oat cultivar (factor II), and study years (factor III).

Soil Type	Years	Oat Cultivar				Mean [1] ± SD [3]
		'Celer'	'Furman'	'Grajcar'	'Kasztan'	
Stagnic Luvisol	2012	48.0	47.3	51.5	49.5	49.1 ± 1.85
	2013	55.6	50.9	58.5	49.2	53.5 ± 4.29
	2014	51.0	54.8	52.8	52.9	52.9 ± 1.55
	Mean [2] ± SD [3]	51.5 ± 3.84	51.0 ± 3.73	54.3 ± 3.74	50.5 ± 2.03	51.8 A
Haplic Cambisol	2012	38.1	38.8	40.0	27.5	36.1 ± 5.80
	2013	51.3	45.9	45.2	46.6	47.3 ± 2.75
	2014	54.0	52.2	54.6	51.4	53.0 ± 1.50
	Mean [2] ± SD [3]	47.8 ± 8.46	45.6 ± 6.71	46.6 ± 7.39	41.8 ± 12.6	45.5 B
Mean	2012	43.1	43.1	45.8	38.5	42.6 ± 3.00 z
	2013	53.4	48.4	51.9	47.9	50.4 ± 2.69 y
	2014	52.5	53.5	53.7	52.1	53.0 ± 0.77 x
	Mean [2] ± SD [3]	49.7 ± 5.74 ab	48.3 ± 5.22 bc	50.4 ± 4.17 a	46.2 ± 6.97 c	48.6

Table 14. *Cont.*

Soil Type	Years	Oat Cultivar				Mean [1] ± SD [3]
		'Celer'	'Furman'	'Grajcar'	'Kasztan'	
LSD $_{0.05}$ soil type				1.67		
LSD $_{0.05}$ cultivar				1.97		
LSD $_{0.05}$ years				1.97		
LSD $_{0.05}$ soil type × cultivar				2.78		
LSD $_{0.05}$ soil type × years				2.54		
LSD $_{0.05}$ cultivar × years				3.52		

[1] Mean for the soil type, regardless of the oat cultivar; [2] Mean for the year 2012–2014; [3] S.D.—standard deviation. Homogeneous groups were created for the main factors. According to Tukey's test, mean values marked with the same letters do not differ significantly ($p \leq 0.05$). Capital letters (A and B) for mean values of the first factor levels—soil types, small letters (a, b, c) for mean values of the second-factor levels—oats cultivars and x, y and z letters for the third-factor levels—study years were chosen. The three-factor ANOVA—first-factor, soil type: Stagnic Luvisol and Haplic Cambisol; second-factor, oat cultivar: 'Celer', 'Furman', 'Grajcar', 'Kasztan'; third-factor, years: 2012, 2013, 2014.

3.4. Protein Content in Oat Grains and Vetch Seeds

The soil type significantly differentiated the total protein content in oat grains (Table 15). A 5% higher protein content was found in grains of oats grown in H.C. soil than the S.L. soil. The protein content differed among the oat cultivars in the mixtures and was in a range of 94.4 for cv. 'Kasztan' to 107 g kg^{-1} for cv. Furman. On average, a 39% higher protein content was found in the grains of oats in 2013 than in the dry 2012 year.

Table 15. Total protein content in oat grain (g kg^{-1}) grown in the mixtures, depending on the soil type (factor I), oat cultivar (factor II), and study years (factor III).

Soil Type	Years	Oat Cultivar				Mean [1] ± SD [3]
		'Celer'	'Furman'	'Grajcar'	'Kasztan'	
Stagnic Luvisol	2012	86.7	70.3	70.0	73.0	75.0 ± 7.95
	2013	120	141	135	106	125 ± 15.8
	2014	87.6	104	113	90.5	98.5 ± 11.6
	Mean [2] ± SD [3]	97.9 ± 18.7	105 ± 35.1	106 ± 33.2	89.8 ± 16.4	99.6 B
Haplic Cambisol	2012	71.4	88.0	80.2	83.6	80.8 ± 7.04
	2013	144	130	130	121	131 ± 9.69
	2014	103	111	101	93.0	102 ± 7.18
	Mean [2] ± SD [3]	106 ± 36.4	110 ± 21.1	104 ± 25.2	99.1 ± 19.2	105.0 A
Mean	2012	79.1	79.1	75.1	78.3	77.9 ± 1.92 z
	2013	132	135	133	113	128.0 ± 10.2 x
	2014	95.0	107	107	91.8	100.0 ± 7.86 y
	Mean [2] ± SD [3]	102 ± 27.0 c	107 ± 28.1 a	105 ± 28.9 b	94.4 ± 17.5 d	102.0
LSD $_{0.05}$ soil type				0.942		
LSD $_{0.05}$ cultivar				1.65		
LSD $_{0.05}$ years				1.13		
LSD $_{0.05}$ soil type × cultivar				2.13		
LSD $_{0.05}$ soil type × years				1.43		
LSD $_{0.05}$ cultivar × years				2.26		

[1] Mean for the soil type, regardless of the oat cultivar; [2] Mean for the year 2012–2014; [3] S.D.—standard deviation. Homogeneous groups were created for the main factors. According to Tukey's test, mean values marked with the same letters do not differ significantly ($p \leq 0.05$). Capital letters (A and B) for mean values of the first factor levels—soil types, small letters (a, b, c) for mean values of the second-factor levels—oats cultivars and x, y and z letters for the third-factor levels—study years were chosen. The three-factor ANOVA—first-factor, soil type: Stagnic Luvisol and Haplic Cambisol; second-factor, oat cultivar: 'Celer', 'Furman', 'Grajcar', 'Kasztan'; third-factor, years: 2012, 2013, 2014.

The type of soil significantly affected the vetch seeds' protein content, which was higher on the S.L. soil (Table 16). High protein content in vetch seeds was found in the mixture with oats cv. Furman, which was also rich in protein. The same relationship was found for the lowest protein content in the vetch/oat mixture, which was in the one with oat cv. 'Kasztan' (Table 15). On average, the highest protein content in vetch seeds was found in 2013, and the lowest in 2014.

Table 16. Total protein content in vetch seeds (g kg^{-1}) grown in the mixtures, depending on the soil type (factor I), oat cultivar (factor II), and study years (factor III).

Soil Type	Years	Oat Cultivar				Mean [1] ± SD [3]
		'Celer'	'Furman'	'Grajcar'	'Kasztan'	
Stagnic Luvisol	2012	273	272	268	274	272 ± 2.50
	2013	292	293	285	274	286 ± 8.80
	2014	286	264	273	272	273 ± 9.12
	Mean [2] ± SD [3]	284 ± 9.59	276 ± 15.4	275 ± 8.89	273 ± 1.37	277 A
Haplic Cambisol	2012	287	290	276	267	280 ± 10.5
	2013	271	273	276	272	273 ± 2.40
	2014	273	275	275	263	271 ± 5.50
	Mean [2] ± SD [3]	277 ± 8.90	279 ± 9.07	276 ± 0.95	267 ± 4.76	275 B
Mean	2012	280	281	272	270	276 ± 5.40 y
	2013	281	283	281	273	280 ± 4.39 x
	2014	279	269	274	267	272 ± 5.24 z
	Mean [2] ± SD [3]	280 ± 1.15 a	278 ± 7.65 b	275 ± 4.66 c	270 ± 2.99 d	276
LSD $_{0.05}$ soil type				0.794		
LSD $_{0.05}$ cultivar				1.33		
LSD $_{0.05}$ years				0.764		
LSD $_{0.05}$ soil type × cultivar				1.72		
LSD $_{0.05}$ soil type × years				1.01		
LSD $_{0.05}$ cultivar × years				1.53		

[1] Mean for the soil type, regardless of the oat cultivar; [2] Mean for the year 2012–2014; [3] S.D.—standard deviation. Homogeneous groups were created for the main factors. According to Tukey's test, mean values marked with the same letters do not differ significantly ($p \leq 0.05$). Capital letters (A and B) for mean values of the first factor levels—soil types, small letters (a, b, c) for mean values of the second-factor levels—oats cultivars and x, y and z letters for the third-factor levels—study years were chosen. The three-factor ANOVA—first-factor, soil type: Stagnic Luvisol and Haplic Cambisol; second-factor, oat cultivar: 'Celer', 'Furman', 'Grajcar', 'Kasztan'; third-factor, years: 2012, 2013, 2014.

The canonical variate analysis (CVA), which included all the tested traits, was applied to extract the factor that influenced the overall state of the oat–vetch mixtures the most (Figure 4). The first two canonical variates explained jointly 81.19% of the total variation between the treatments. The greatest, significant linear relationship was found for protein content in oat grains (g kg^{-1}) and a share of common vetch seed in the mixture's yield (positive dependency). The significant negative dependencies were found for the mixtures' yield, the number of oats panicles per m^{-2}, and the mass of grains per oat panicle. The second canonical variate was significantly positively correlated with the number of oat panicles per m^{-2} and the density of oat at spring. The negative correlation was found for the number of grains per oat panicle, a share of vetch seed in the mixture's yield, and the 1000-grain mass of oat.

The diversities in all traits, as measured with Mahalanobis distances, are presented in Table 17.

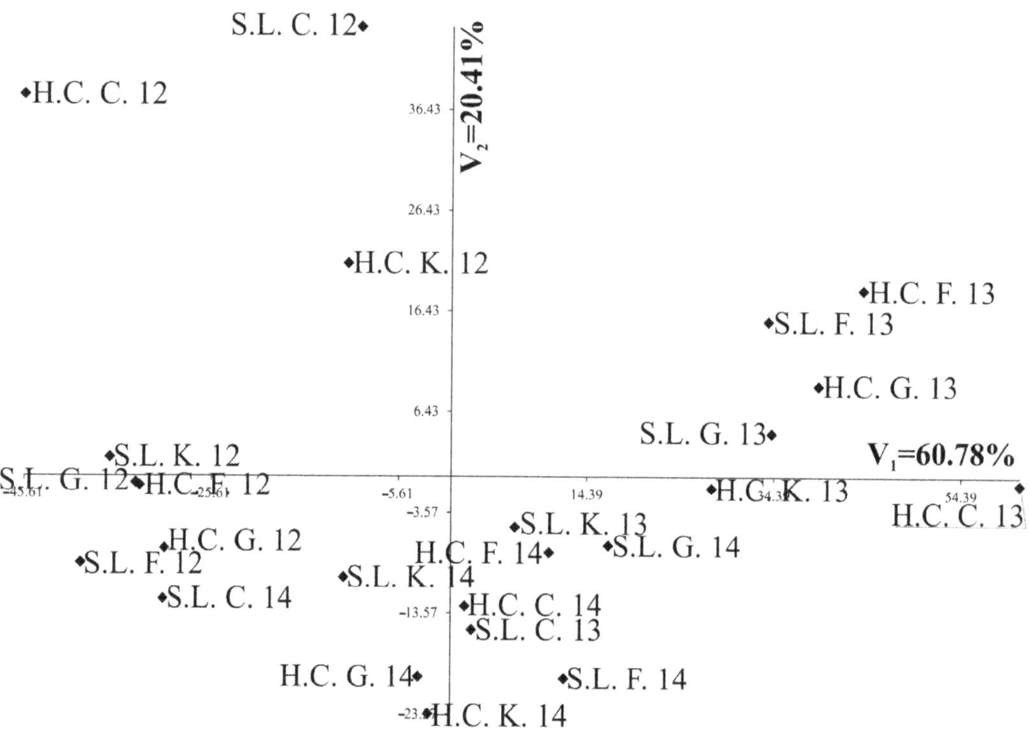

Figure 4. Distribution of combinations of treatments in the two first canonical variates. Abbreviations: S.L.—Stagnic Luvisolor, H.C.—Haplic Cambisol; C.—'Celer', F.—'Furman', G.—'Grajcar', K.—'Kasztan'; 12–14—years 2012–2014.

The CVA analysis pointed to the year (weather conditions) as a main differentiating factor for the mixture's performance. The best for the mixtures turned to be the year 2013, and the worst—the dry year 2012. Moreover, Haplic Cambisol was better for the tested mixtures than the Stagnic Luvisol. The analysis also revealed that among the studied four cultivars of oats, the best for mixing with vetch cv. 'Hanka' was cv. 'Furman' and cv. 'Grajcar'.

Table 17. Mahalanobis distances between pairs of combinations of three studied factors.

		1	2	3	4	5	6	7	8	9	10	11	12	13	14	15	16	17	18	19	20	21	22	23	24
S.L. C. 12	1																								
S.L. F. 12	2	62.07																							
S.L. G. 12	3	52.52	15.57																						
S.L. K. 12	4	51.55	12.52	13.09																					
S.L. C. 13	5	74.73	58.29	61.56	57.87																				
S.L. F. 13	6	64.92	84.78	81.65	79.04	47.67																			
S.L. G. 13	7	65.65	79.34	75.17	74.84	43.55	19.67																		
S.L. K. 13	8	54.44	49.02	45.12	46.52	35.24	45.76	34.66																	
S.L. C. 14	9	67.78	28.59	36.68	30.63	39.32	72.16	68.22	42.72																
S.L. F. 14	10	68.71	54.86	50.72	55.09	49.38	59.18	45.38	24.24	55.55															
S.L. G. 14	11	59.39	57.19	52.19	55.12	44.35	44.57	31.68	17.36	54.54	19.64														
S.L. K. 14	12	55.76	30.51	27.97	30.28	41.73	61.61	52.52	21.11	30.37	28.56	29.55													
H.C. C. 12	13	45.78	53.41	49.69	44.58	76.33	87.73	88.76	69.53	55.71	87.47	80.56	63.36												
H.C. F. 12	14	61.66	32.12	35.01	30.03	44.76	72.9	69.54	47	21.96	61.71	57.66	37.02	43.84											
H.C. G. 12	15	58.94	18.18	19.3	18.55	48.87	76.41	69.85	40.25	26.43	49.3	49.99	26.24	49.86	22.08										
H.C. K. 12	16	28.23	45.02	34.01	36.47	65.09	65.62	60.9	38.97	55.23	48.63	43.69	37.06	50.18	50.26	39.95									
H.C. C. 13	17	85.57	101.8	96	98.62	72.08	51.26	40.07	56.48	97.29	56.12	48.11	75.27	114.73	98.69	92.27	77.35								
H.C. F. 13	18	61.33	88.71	80.96	83.29	67.64	39.77	31.83	45.53	85.42	52.33	38.51	63.4	93.67	84.3	79.66	57.2	28.28							
H.C. G. 13	19	62.9	81.76	75.09	77.33	58.73	34.73	23.68	37.6	77	43.68	28.34	55.28	91.66	76.72	72.66	55.4	29.34	13.33						
H.C. K. 13	20	61.53	69.1	63.64	65.78	48.74	41.88	28.82	25.5	65.86	31.29	20.03	43.58	84.94	67.16	59.51	48.09	33.6	26.95	20.51					
H.C. C. 14	21	61.11	45.09	42.17	43.9	35.87	55.92	44.2	15.88	41.73	24.11	25.27	22.41	70.72	44.79	34.02	41.66	61.2	54.34	46.69	30.15				
H.C. F. 14	22	58.91	53.88	49.11	50.69	40.18	46.87	34.82	17.11	49.56	23.02	18.48	28.49	74.89	50.41	42.99	41.07	54.26	45.05	36.32	24.67	16.51			
H.C. G. 14	23	66.97	41.07	39.81	42.19	35.95	59.91	48.16	20.21	35.44	24.05	26.43	16.49	73.23	41.07	32.14	47.18	69.02	61.79	52.08	38.84	15.54	22.13		
H.C. K. 14	24	70.26	43.76	41.11	45.91	50.75	70.85	58.48	29.34	48.57	20.45	31.58	25.96	80.34	53.81	37.17	46.31	68.47	63.88	56.15	39.82	19.66	27.66	20.89	

Abbreviations: S.L.—Stagnic Luvisol, H.C.—Haplic Cambisol; C.—'Celer', F.—'Furman', G.—'Grajcar', K.—'Kasztan'; 12–14—years 2012–2014.

4. Discussion

Our results clearly show that the weather course during the vegetation season is a primary factor affecting the performance of oat–vetch mixtures. Interestingly, oats had higher yields during the dry season, whereas vetch had higher yields during seasons classified as regular. Many authors emphasize that oats are more competitive toward companion species in mixtures during dry seasons [32–37]. In adverse weather conditions, such as rainfall shortage or inadequate rainfall distribution during vegetation and lack of radiation, the cereal component determines the cereal–legume mixture's yield [38].

We also showed that particular components of the mixture preferred different soil types; oat yielded better on a fertile Stagnic Luvisol. The vetch's yield parameters were better on a sandy Haplic Cambisol of a low N content. Moreover, vetch was performing better than oat in the following years of the experiment, when the rainfall distribution was variable. The balance between species is a key factor determining productivity of mixtures [39–41]. One of the management factors that affect intercropped species' relative competitiveness and performance is N availability [35–37,42–44]. According to [45], the yield of cereal-legume mixtures grown on the poorer soils depends mainly on the cereal component and the species and sowing density of lupine do not have a significant impact on the yield of mixtures.

On the other hand, [44] showed that the mixtures of oats with yellow lupine and triticale with lupine yielded the best on the soil intended for rye cultivation. Other authors [42,46,47] underline that the legume component performs better in a situation of N deficiency, which may happen in the organic crop rotations. Cultivation of mixtures of oat and legumes is beneficial through the structure-forming action of the legume root system [48], increasing soil biodiversity and activating nutrients from compounds inaccessible to the root system of cereals [49].

In our research, oat cultivars significantly differentiated the yield, total protein content of oat grains and vetch seeds, as well as the vetch yield parameters, such as the 1000-seed mass. Similar results were obtained by [50], who found that in the oat–vetch mixture grown for fodder, the selection of oats cultivars and vetch species affected the crude protein content in the mixtures' biomass. However, [37] underlines that the N content in pea grain is lower in the mixture with cereal, compared to the pure sowing. The reverse situation was noted for the mixture's cereal component, in which N content in both grain and straw was higher [37].

Cultivation of crop mixtures, composed of at least two species, is a crucial element of proper agricultural technology, particularly in conventional farming, but mostly in the organic one [42,51,52]. The results of our study showed that the yielding and protein content of interspecies mixtures is the result of many natural and agrotechnical factors [7]. Therefore, the identification of yield variability of legume-cereal mixtures is particularly important and justified due to climate change, and more frequently occurring water shortage, as they are considered an important element of agricultural diversification [53].

5. Conclusions

The course of the weather in particular years was the main factor affecting the performance of the organically grown oat–vetch mixture. In warm and dry weather oat component of the mixture affected the final yield. Among the oat cultivars in a mixture with common vetch, the 'Kasztan' cultivar was characterized by the highest yield, but varied over the years. In a dry and very warm year with low variability of rainfall distribution, it yielded the highest. On the other hand, warm and average years, with a high variability of rainfall distribution, presented the lowest yields compared to other cultivars.

Common vetch grown with oat increased the protein content of the oat grain. The highest content of total protein was measured in grains of cv. Furman. On the other hand, the highest content of total protein in vetch seeds was in cultivation with 'Celer' cultivar. The highest share of vetch seed in the grain yield of the mixture was noted with cv. 'Grajcar'.

The type of soil was also crucial. A higher yield of mixture was found on Stagnic Luvisol soil whereas total protein content was higher in the mixture grown on Haplic Cambisol soil. On Stagnic Luvisol, the Furman cultivar performed better whereas 'Kasztan' fared better on Haplic Cambisol.

Proper selection of oat cultivar for the mixture with common vetch in conditions of organic farming is an important measure affecting the grain yield, yield parameters, and protein content in vetch seeds.

Author Contributions: Conceptualization, K.P. and S.P.; Methodology, K.P. and S.P.; Validation, A.L.; Formal analysis, K.P.; Investigation, K.P. and S.P.; Data curation, K.P., S.P., and J.B.; Writing—original draft preparation, K.P., S.P. and A.S.; Writing—review and editing, K.P., S.P., J.B. and A.L.; Visualization, K.P., S.P., A.S. and J.B.; Funding acquisition, K.P. and A.L. All authors have read and agreed to the published version of the manuscript.

Funding: This research was financed by the Ministry of Science and Higher Education of the Republic of Poland–partially, this research was funded by the Ministry of Science and Higher Education in Poland for education in the years 2010–2015 as a research project (grant no. N N310 446938).

Institutional Review Board Statement: Not applicable.

Informed Consent Statement: Not applicable.

Data Availability Statement: The data presented in this study are available on request from the corresponding authors.

Conflicts of Interest: The authors declare no conflict of interest.

References

1. Salama, H.S.A. Mixture cropping of berseem clover with cereals to improve forage yield and quality under irrigated conditions of the Mediterranean Basin. *Ann. Agric. Sci.* **2020**, *65*, 159–167. [CrossRef]
2. Kamalongo, D.M.; Cannon, N.D. Advantages of bi-cropping field beans (*Vicia faba*) and wheat (*Triticum aestivum*) on cereal forage yield and quality. *Biol. Agric. Hortic.* **2020**, *36*, 213–229. [CrossRef]
3. Hammelehle, A.; Oberson, A.; Lüscher, A.; Mäder, P.; Mayer, J. Above-and belowground nitrogen distribution of a red clover-perennial ryegrass sward along a soil nutrient availability gradient established by organic and conventional cropping systems. *Plant Soil* **2018**, *425*, 507–525. [CrossRef]
4. Peoples, M.B.; Brockwell, J.; Herridge, D.F.; Rochester, I.J.; Alves, B.J.R.; Urquiaga, S.; Boddey, R.M.; Dakora, F.D.; Bhattarai, S.; Maskey, S.L.; et al. The contributions of nitrogen-fixing crop legumes to the productivity of agricultural systems. *Symbiosis* **2009**, *48*, 1–17. [CrossRef]
5. Jensen, E.S.; Carlsson, G.; Hauggaard-Nielsen, H. Intercropping of grain legumes and cereals improves the use of soil N resources and reduces the requirement for synthetic fertilizer N: A global-scale analysis. *Agron. Sustain. Dev.* **2020**, *40*, 5. [CrossRef]
6. Mielniczuk, E.; Patkowska, E.; Jamiołkowska, A. The influence of catch crops on fungal diversity in the soil and health of oat. *Plant Soil Environ.* **2020**, *66*, 99–104. [CrossRef]
7. Klima, K.; Synowiec, A.; Puła, J.; Chowaniak, M.; Pużyńska, K.; Gala-Czekaj, D.; Kliszcz, A.; Galbas, P.; Jop, B.; Dąbkowska, T.; et al. Long-term productive, competitive, and economic aspects of spring cereal mixtures in integrated and organic crop rotations. *Agriculture* **2020**, *10*, 231. [CrossRef]
8. Sadiq Butt, M.; Tahir-Nadeem, M.; Khan, M.K.I.; Shabir, R.; Butt, M.S. Oat: Unique among the cereals. *Eur. J. Nutr.* **2008**, *47*, 68–79. [CrossRef]
9. Gambuś, H.; Gibiński, M.; Pastuszka, D.; Mickowska, B.; Ziobro, R.; Witkowicz, R. The application of residual oats flour in bread production in order to improve its quality and biological value of protein. *Acta Sci. Pol. Techn. Aliment.* **2011**, *10*, 317–325.
10. Biel, W.; Kazimierska, K.; Bashutska, U. Nutritional value of wheat, triticale, barley and oat grains. *Acta Sci. Pol. Zootech.* **2020**, *19*, 19–28. [CrossRef]
11. Welch, R.W.; Hayward, M.V.; Jones, D.I.H. The composition of oat husk and its variation due to genetic and other factors. *J. Sci. Food Agric.* **1983**, *34*, 417–426. [CrossRef]
12. Lawrence, T.L.J. The effects of cereal particle size and pelleting on the nutritive value of oat-based diets for the growing pig. *Anim. Feed Sci. Technol.* **1983**, *8*, 91–97. [CrossRef]
13. Eskandari, H.; Ghanbari, A.; Javanmard, A. Intercropping of cereals and legumes for forage production. *Not. Sci. Biol.* **2009**, *1*, 7–13. [CrossRef]
14. Lauk, R.; Lauk, E. Pea-oat intercrops are superior to pea-wheat and pea-barley intercrops. *Acta Agric. Scand. B Soil Plant Sci.* **2008**, *58*, 139–144. [CrossRef]
15. Rebolé, A.; Alzueta, C.; Ortiz, L.T.; Barro, C.; Rodríguez, M.L.; Caballero, R. Yields and chemical composition of different parts of the common vetch of flowering and at two seed filling states. *Span. J. Agric. Res.* **2004**, *2*, 550. [CrossRef]

16. Sadeghi, G.H.; Tabeidian, S.A.; Toghyani, M. Effect of processing on the nutritional value of common vetch (*Vicia sativa*) Seed as a feed ingredient for broilers. *J. Appl. Poult. Res.* **2011**, *20*, 498–505. [CrossRef]
17. Ceglarek, F.; Rudziński, R.; Buraczyńska, D. The effect of the amount of seeds sown on the crop structure elements and seed yields of common vetch grown as pure and mixed crops with supporting plants. *Ann. UMCS* **2004**, *59*, 1147–1154.
18. Vasileva, V.; Kosev, V.; Kaya, Y. Evaluation of winter vetch varieties by quality indicators. *Int. J. Innov. Approaches Agric. Res.* **2020**, *4*, 156–165.
19. Vázquez, E.; Benito, M.; Espejo, R.; Teutscherova, N. No-tillage and liming increase the root mycorrhizal colonization, plant biomass and N content of a mixed oat and vetch crop. *Soil Tillage Res.* **2020**, *200*, 104623. [CrossRef]
20. Polishchuk, V.; Zuravel, S.; Kravchuk, M.; Klymenko, T. Organic technology of growing vetch and oat mixture under condition of using organic and mineral preparations under different fertilization systems. *Sci. Eur.* **2020**, *47*, 4–12.
21. Alemu, B.; Melaku, S.; Prasad, N.K. Effects of varying seed proportions and harvesting stages on biological compatibility and forage yield of oats (*Avena sativa* L.) and vetch (*Vicia villosa* R.) mixtures. *Livestock Res. Rural Dev.* **2007**, *19*, 12.
22. Dolata, A.; Andrzejewska, J.; Wiatr, K. Reaction of determinate and indeterminate common vetch (*Vicia sativa* L. ssp. *sativa*) cultivars to different climatic and soil conditions. *Acta Sci. Pol. Agric.* **2006**, *5*, 25–35.
23. IUSS Working Group. World Reference Base for Soil Resources. In *World Soil Resources Report No. 103*; FAO: Rome, Italy, 2006.
24. Beral, A.; Rincent, R.; Le Gouis, J.; Girousse, C.; Allard, V. Wheat individual grain-size variance originates from crop development and from specific genetic determinism. *PLoS ONE* **2020**, *15*, e0230689. [CrossRef]
25. Shapiro, S.S.; Wilk, M.B. An analysis of variance test for normality (complete samples). *Biometrika* **1965**, *52*, 591–611. [CrossRef]
26. Rencher, A.C. Interpretation of canonical discriminant functions, canonical variates, and principal components. *Am. Stat.* **1992**, *46*, 217–225.
27. Seidler-Łożykowska, K.; Bocianowski, J. Evaluation of variability of morphological traits of selected caraway (*Carum carvi* L.) genotypes. *Ind. Crops Prod.* **2012**, *35*, 140–145. [CrossRef]
28. Mahalanobis, P.C. On the generalized distance in statistics. *Proc. Natl. Acad. Sci. India* **1936**, *12*, 49–55.
29. Dzieżyc, J. *Czynniki Plonotwórcze Plonowanie Roślin*; PWN: Warszawa, Poland, 1993; pp. 1–474.
30. Kaczorowska, Z. *Rainfall in Poland Over a Long-Term Cross-Section*; Geographic Works; Institute of Geography, Polish Academy of Sciences: Warsaw, Poland, 1962.
31. Ziernicka, A. Klasyfikacja odchyleń temperatury powietrza od normy w Polsce południowo-wschodniej. *Zeszyty Naukowe Akademii Rolniczej Krakówie* **2001**, *22*, 5–18.
32. Szpunar-Krok, E.; Bobrecka-Jamro, D.; Tobiasz-Salach, R. Yielding of naked oat and faba bean in pure sowing and mixtures. *Fragm. Agron.* **2009**, *26*, 145–151.
33. Pisulewska, E.; Witkowicz, R.; Kidacka, A. Yield, yield components, and accuracy of grain in selected cultivars of oats. *Żywność Nauka Technologia Jakość* **2010**, *3*, 117–126.
34. Klima, K.; Stokłosa, A.; Pużyńska, K. Agricultural and economic circumstances of cereal cultivation under differentiated soil and climate conditions. *Zeszyty Problemowe Postepow Nauk Rolniczych* **2011**, *559*, 115–121.
35. Iqbal, M.A.; Hamid, A.; Ahmad, T.; Siddiqui, M.H.; Hussain, I.; Ali, S.; Ali, A.; Ahmad, Z. Forage sorghum-legumes intercropping: Effect on growth, yields, nutritional quality and economic returns. *Bragantia* **2019**, *78*, 82–95. [CrossRef]
36. Iqbal, M.A.; Hamid, A.; Hussain, I.; Siddiqui, M.H.; Ahmad, T.; Khaliq, A.; Ahmad, Z. Competitive indices in cereal and legume mixtures in a South Asian environment. *Agron. J.* **2019**, *111*, 242–249. [CrossRef]
37. Monti, M.; Pellicanò, A.; Santonoceto, C.; Preiti, G.; Pristeri, A. Yield components and nitrogen use in cereal-pea intercrops in mediterranean environment. *Field Crops Res.* **2016**, *196*, 379–388. [CrossRef]
38. Layek, J.; Das, A.; Mitran, T.; Nath, C.; Meena, R.S.; Yadav, G.S.; Shivakumar, B.G.; Kumar, S.; Lal, R. Cereal + legume intercropping: An option for improving productivity and sustaining soil health. In *Legumes for Soil Health and Sustainable Management*; Springer: Singapore, 2018; pp. 347–386.
39. Rudnicki, F.; Wenda-Piesik, A. Productivity of pea-cereal intercrops on good rye soil complex. *Zeszyty Problemowe Postepow Nauk Rolniczych* **2007**, *516*, 181–195.
40. Wenda-Piesik, A.; Rudnicki, F. The importance of pea cultivar selection for pea-cereal intercropping on good rye soil complex. *Zeszyty Problemowe Postepow Nauk Rolniczych* **2007**, *516*, 277–292.
41. Rudnicki, F.; Kotwica, K. Competitive interactions between spring cerealsand lupinsin mixtures and mixture growing production effectson very good rye complex soil. *Fragm. Agron.* **2007**, *4*, 145–152.
42. Yu, Y.; Stomph, T.-J.; Makowski, D.; Zhang, L.; van der Werf, W. A meta-analysis of relative crop yields in cereal/legume mixtures suggests options for management. *Field Crops Res.* **2016**, *198*, 269–279. [CrossRef]
43. Gaudio, N.; Escobar-Gutiérrez, A.J.; Casadebaig, P.; Evers, J.B.; Gerard, F.; Louarn, G.; Colbach, N.; Munz, S.; Launay, M.; Marrous, H.; et al. Current knowledge and future research opportunities for modeling annual crop mixtures. A review. *Agron. Sustain. Dev* **2019**, *39*, 20. [CrossRef]
44. Kotwica, K.; Rudnicki, F. Production effects of growing spring cereal and cereal-and-legume mixtures on good rye complex soil. *Acta Sci. Pol. Agric.* **2004**, *3*, 149–156.
45. Kotwica, K.; Rudnicki, F. Composing of spring cereal mixtures with lupine on light soil. *Zeszyty Problemowe Postepow Nauk Rolniczych* **2003**, *495*, 163–170.

46. Rodriguez, C.; Carlsson, G.; Englund, J.-E.; Flöhr, A.; Pelzer, E.; Jeuffroy, M.-H.; Makowski, D.; Jensen, E.S. Grain legume-cereal intercropping enhances the use of soil-derived and biologically fixed nitrogen in temperate agroecosystems. A meta-analysis. *Eur. J. Agron.* **2020**, *118*, 126077. [CrossRef]

47. Księżak, J. The development of pea and spring barley plants in the mixtures on various soil types. *Zeszyty Problemowe Postepow Nauk Rolniczych* **2007**, *516*, 83–90.

48. Głąb, T.; Pużyńska, K.; Pużyński, S.; Palmowska, J.; Kowalik, K. Effect of organic farming on a stagnic luvisol soil physical quality. *Geoderma* **2016**, *282*, 16–25. [CrossRef]

49. Gentsch, N.; Boy, J.; Guggenberger, G. Incorporation of diverse catch crop mixtures in crop rotation cycles increase biodiversity and nutrient availability in soils. In *Jahrestagung der DBG*; Horizonte des Bodens: Göttingen, Germany, 2017.

50. Molla, E.A.; Wondimagegn, B.A.; Chekol, Y.M. Evaluation of biomass yield and nutritional quality of oats–vetch mixtures at different harvesting stage under residual moisture in Fogera district, Ethiopia. *Agric. Food Secur.* **2018**, *7*. [CrossRef]

51. Pużyńska, K.; Synowiec, A.; Pużyński, S.; Bocianowski, J.; Klima, K.; Lepiarczyk, A. Selected canopy indices and yielding of oats-vetch mixture in two contrasting farming systems. *Agriculture* **2021**, in press.

52. Kaut, A.H.E.E.; Mason, H.E.; Navabi, A.; O'Donovan, J.T.; Spaner, D. Organic and conventional management of mixtures of wheat and spring cereals. *Agron. Sustain. Dev.* **2008**, *28*, 363–371. [CrossRef]

53. Paut, R.; Sabatier, R.; Tchamitchian, M. Modelling crop diversification and association effects in agricultural systems. *Agric. Ecosyst. Environ.* **2020**, *288*, 106711. [CrossRef]

agriculture

MDPI

Article

The Performance of Oat-Vetch Mixtures in Organic and Conventional Farming Systems

Katarzyna Pużyńska [1],*, Agnieszka Synowiec [1], Stanisław Pużyński [2],*, Jan Bocianowski [3], Kazimierz Klima [1] and Andrzej Lepiarczyk [1]

1   Department of Agroecology and Crop Production, Faculty of Agriculture and Economics, University of Agriculture in Krakow, Mickiewicza 21, 31-120 Krakow, Poland; agnieszka.synowiec@urk.edu.pl (A.S.); kazimierz.klima@urk.edu.pl (K.K.); andrzej.lepiarczyk@urk.edu.pl (A.L.)
2   Malopolska Agricultural Advisory Centre, Osiedlowa 9, 32-082 Karniowice, Poland
3   Department of Mathematical and Statistical Methods, Poznań University of Life Sciences, Wojska Polskiego 28, 60-637 Poznań, Poland; jan.bocianowski@up.poznan.pl
*   Correspondence: katarzyna.puzynska@urk.edu.pl (K.P.); spuzynski@wp.pl (S.P.); Tel.: +48-12-662-4368 (K.P.)

Abstract: The research aimed to compare the yields and yield components of mixtures of oats with common vetch grown for seeds in organic and conventional farming systems. Moreover, the selection of oat cultivars for the mixture and its performance in a crop rotation experiment in different growing years was analyzed. Additionally, the leaf area index (LAI) and the relative content of chlorophyll (SPAD) of the mixtures were assessed. The field experiment with four-field crop rotation in organic or conventional farming systems was carried out in 2012–2014 in southern Poland. Common vetch (*Vicia sativa* L., cv. 'Hanka') was mixed with one of two oat (*Avena sativa* L.) cultivars, 'Celer' or 'Grajcar.' The effects of all of the factors on the mixtures' canopy indices and yield were found. The canonical analysis revealed that the weather course, especially drought, had the largest effect on the oat-vetch mixtures' performance. Moreover, the mixtures developed the highest LAI (5.28 $m^2 \cdot m^{-2}$) and seed yield (4.57 t ha^{-1}) in the conventional farming system. On the contrary, the share of vetch seeds in the mixtures was 24% higher in the organic system than in the conventional one. The selection of cv. 'Grajcar' oats for the mixture with vetch increased the share of vetch seeds in the yield by 16.5%. In summary, a balanced share of oat-vetch mixture components depends on the proper selection of the oat cultivar, especially for organic farming systems.

Keywords: cereal-legume mixture; organic farming; conventional farming; leaf area index; leaf greenness index; seed yield; yield components

Citation: Pużyńska, K.; Synowiec, A.; Pużyński, S.; Bocianowski, J.; Klima, K.; Lepiarczyk, A. The Performance of Oat-Vetch Mixtures in Organic and Conventional Farming Systems. *Agriculture* **2021**, *11*, 332. https://doi.org/10.3390/agriculture11040332

Academic Editors: Urs Feller and Les Copeland

Received: 19 January 2021
Accepted: 3 April 2021
Published: 8 April 2021

Publisher's Note: MDPI stays neutral with regard to jurisdictional claims in published maps and institutional affiliations.

1. Introduction

In Europe, cereal-legume mixtures have long been considered minor crops. However, interest in their cultivation has been growing in recent years, as they are considered an important element of agricultural diversification [1]. For example, in Poland, in 2019, the cultivation area of cereal-legume mixtures was 0.27% (29,300 ha) of the total arable land, of which the majority were spring mixtures [2]. The mixtures are cultivated in organic and sustainable agricultural systems [3,4]; they are cultivated mainly for high-protein fodder, green fodder, hay, or green manure [5–8].

Crop mixtures are essential for crop rotations in organic farming [9–11], contributing to several ecosystem services [12]; they are responsible for the maintenance of greater species diversity in crop-rotation [13,14], an increase in biologically bound nitrogen in soil [15,16], and a decrease in disease and pest outcomes [17]. Moreover, cereal-legume mixtures with varying rooting depth improve soil structure, i.e., by loosening deeper layers of soil [18,19], making mechanical operations easier. Contrarily, in conventional farming, which is cash-oriented, the role of cereal-legume mixtures is marginal. That is because mineral fertilizers and pesticides replace the mixtures' nutritional and pesticide properties. There has been a trend in agriculture in recent years to shift from traditional conventional farming to sustainable, more environmentally friendly farming, increasing the inclusion of these mixtures in crop rotation [20].

One of the spring cereal-legume mixtures, relatively popular in cultivation in temperate climate, is oats with common vetch [21]. Both components of this mixture differ in soil and climatic requirements and agrotechnology. They offer a premise for an appropriate selection of species and cultivars, and proportions of mixture, for sowing [22]. According to many authors, crop mixtures' yielding depends on the proper selection of cultivars [13,23–25]. Common vetch is a valuable component of these mixtures due to the high protein content of its seeds. However, vetch grown in a mixture with oats is characterized by little competitive potential, especially for light [26]. This translates into a lower growth of vetch that develops smaller seeds of lower nutrient content.

On the other hand, even though a highly competitive species in mixtures [27], oats support the companion crop from lodging [28]. The maximum demand for water and nutrients of both mixture components elapses during the vegetation. For that reason, interspecific competition in mixtures is lower than in the case of intraspecific competition in pure sowing [20].

Several indices measure the condition of the crop canopy, i.e., the leaf area index (LAI) and the leaf chlorophyll content (SPAD). The LAI informs about the leaves' area, which is equal to the assimilation area [29]. On the other hand, the SPAD shows the relative content of chlorophyll in the leaves, translating into their nitrogen nutrition [30]. As a result, there is a relationship between the LAI and SPAD values and the seed yield [13,31,32]. Klima et al. [13] correlate higher values of LAI of spring cereals mixtures with higher mixtures' yields than their pure sowings. However, the LAI of the oat-vetch mixture has not been studied so far.

The main aim of the study was to compare the yield and yield components of mixtures of oats (*Avena sativa* L.) with common vetch (*Vicia sativa* L.) in two farming systems differing in fertilization and plant protection means. The selection of oat cultivars on the yield of mixtures, including temperature and rainfall during 2012–2014, was also analyzed. Additionally, the leaf area index (LAI) of mixtures and the relative chlorophyll content (SPAD) in oats and vetch leaves were measured in two phases of plants' growth.

2. Materials and Methods

2.1. Field Site and Experiment Descriptions

The four-field crop rotation: potato—winter wheat—oats and common vetch mixture—winter spelt, in a randomized split-split-plot design, has been carried out since 2009 in the Experimental Station Mydlniki-Kraków, Poland (50°04′ N, 19°51′ E, 280 m a.s.l., Figure 1), on Stagnic Luvisol (SL) soil [33]. All crops were present each year, which means that the mixture of oats and vetch was sown every year following the winter wheat.

The investigations for this paper were carried out in the years 2012–2014. The examined soil texture was loam developed from loess; pH (KCl) 6.04; N_{tot} 0.858 g kg^{-1}; P 423.2 mg kg^{-1} soil; K 148.2 mg kg^{-1} soil; and C_{org} 7.34%.

The first factor of the experiment was the farming system: (i) organic—without any artificial mean; and (ii) conventional with synthetic pesticides and mineral fertilizers. The second factor was selecting the oat cultivars: 'Celer' or 'Grajcar' for the mixture with common vetch cv. 'Hanka'. The course of temperature and precipitation in 2012–2014 was considered the third factor.

The oat and vetch mixtures were sown at the optimal agrotechnical dates for southern Poland, 23 March 2012; 16 April 2013; and 20 March 2014, at a planned density of plants per m^{-2} 500 and 75 for the oats and vetch, respectively. The mixtures were sown on plots of 24 m^2 area (3 × 8 m), using a plot drill (Hege 80) at a row space of 13.0 cm. A total of 16 plots were present each year (four replications for every mixture in both systems). Soil tillage was similar in organic and conventional plots. It consisted of a deep pre-winter plowing (October) and shallow seedbed tillage using an active harrow and a string roller (April).

Figure 1. Location of the study site. Source: https://www.google.com/maps/ (Accessed on 17 January 2021).

Every four years, 30 tons of composted manure per hectare was used under potato in the conventional and organic system. A mineral fertilization (kg ha^{-1}) of 80 N, 65 P, and 100 K was applied only in the conventional plots. The doses of fertilizers followed good agricultural practices and generally accepted principles of spring cereal cultivation. Nitrogen was applied as ammonium nitrate (34% N); one-third of the dose administered before sowing, and two-thirds as a top dressing. The potassium salt (60% K_2O) and triple superphosphate (40% P_2O_5) were used in full doses before pre-winter plowing in October.

Additionally, in the conventional plots only, three-stage fungicide protection combined with pest control program was applied. Treatments were performed when the economic threshold of pests was exceeded, with a ca. one-month intervals between them. The following pesticides were used: fungicides—prochloraz + tebuconazole or thiophanate-methyl + conazole; insecticides—deltamethrin; beta-cyfluthrin or chlorpyrifos.

In the organic plots, only a mechanic weed control was performed each year by a Weeder harrow at the end of oats' tillering and a manual weed removal before the mixture's harvest.

2.2. Description of Cultivars

According to the breeders' recommendations, the crop cultivars selected for this study are intended to cultivate mountainous areas of temperate climate, where they yield well.

The yellow-grained oat cv. 'Celer' has a 120 days to ripening phase BBCH 85 (German "Biologische Bundesanstalt, Bundessortenamt, und Chemische Industrie") from sowing. The mass of a thousand grains is 41.0 g. The grains have a relatively high proportion of husk (28.2%). The protein and fat content of the grains are 6 (on a 9-point scale, where 9 means most favorable, 5—average 1—least favorable content). Plants are resistant to coronary and stem rust and of good resistance to other diseases. The cv. Celer is relatively short (90 cm), with high lodging resistance. The advised sowing rate of seeds is 550–600 m^{-2}. Breeder: Małopolska Hodowla Roślin (HR), Sp. z o. o., Poland.

The oat cv.'Grajcar' is an early sown cultivar of medium-early ripening, equal to 120 days to the ripening phase (BBCH 85) from sowing. It is a yellow-grained oat, with an average thousand-seed mass of (35.3 g). The grains have a relatively high proportion of husk (29.5%). The protein and fat content are 6 and 7, respectively. The plants are highly resistant to coronary and stem rust. It has average soil requirements. The plants are relatively short (89 cm). The advised sowing rate of seeds is 550–600 seeds m^{-2}. Breeder: Małopolska Hodowla Roślin (HR), Sp. z o. o., Poland.

'Hanka' is common vetch (*Vicia sativa* L.) cultivar of a traditional type of growth, i.e., not self-ending. Plants are lush, 50–160 cm high, rich in leaves ending with sticking tendrils; seeds are brown. The cultivar is very fertile, with seeds of high protein (32%) and low tannins (0.05%). Seeds are ready for harvest 120 days after sowing. The thousand-grain mass is 52 g. It can be grown for seeds, green fodder, or green manure. The cultivar is appropriate for mixing with cereals. Breeder: Firma Nasienna Granum, Poland.

2.3. Leaf Area Index and Leaf Greenness Index

Two indexes of a canopy condition were measured. First, the leaf area index (LAI), characterizing the leaf assimilation area capable of absorbing photosynthetically active radiation (400–700 nm), using a SunScan Canopy Analysis System—SS1-COM Complete System (SunScan Canopy Analysis System, Delta-T Devices Ltd., Burwell, Cambridge, UK). Second, the leaf relative chlorophyll content in soil plant analysis development values (SPAD), using a 502DL chlorophyll meter (Minolta SPAD-502DL, Spectrum Technologies Inc., Plainfield, IL, USA).

The following formulas were used for the calculation of the LAI index (Equations (1) and (2)):

$$K(x,\theta) := \frac{\sqrt{x^2 + \tan(\theta)^2}}{x + 1.702(x + 1.12)^{-0.708}} \tag{1}$$

where:

x is the ellipsoidal leaf angle distribution parameter (ELADP),
θ is the zenith angle of the direct beam,

$$\tau(x,\theta) := \exp(-K(x,\theta)L) \tag{2}$$

where:

τ is the gap fraction,
L is the leaf area index,
$K(x,\theta)$ is the extinction coefficient.

The measurement of relative chlorophyll content by the chlorophyll meter was according to the formula (Equation (3)):

$$M = k \, \log_{10} \frac{I_{0(650)} I_{(940)}}{I_{(650)} I_{0(940)}} \tag{3}$$

where:

k is a confidential proportionality coefficient = 40;
$I_{0(650)}$ is the intensity of incident monochromatic light at 650 nm wavelength;
$I_{(940)}$ is the intensity of transmitted light at 940 nm wavelength;
$I_{(650)}$ is the intensity of transmitted light at 650 nm wavelength;
$I_{0(940)}$ is the intensity of incident monochromatic light at 940 nm wavelength.

The LAI and the SPAD measurements were performed each year on two dates, i.e., LAI_1 and $SPAD_1$ in the oats' tillering phase (BBCH 29), and LAI_2 $SPAD_2$ in the grain watery ripe phase (BBCH 71). The SPAD measurements were performed separately for oats and vetch plants, while the LAI were measured for the mixtures' canopy at four random spots per plot. The SPAD was measured on leaves of 25 plants of oats and vetch per plot.

For the measurement, only fully developed leaves were chosen. The oat's SPAD readings were taken from the middle part of the leaf blade; for vetch this area was the middle leaflet on the pinnate leaf.

2.4. Yield Measurements

Before harvesting, the oat and vetch plants were sampled to determine the number of oat panicles per m^{-2} and grains per panicle, and the number of vetch pods and seeds per pod. The plants were sampled from four random spots of 0.125 m^{-2} each (0.25 m × 0.5 m) across each plot, but three edge rows on both plot sides were omitted. All sampled plants were analyzed, and the results were recalculated to a 1 m^2 area.

The harvest was carried out with a plot harvester (Seedmaster, Wintersteiger) at the oats' fully ripe growth stage (BBCH 97). After the harvest, the oats' grain and vetch seeds from each plot (24 m^2) were weighed. Additional samples of grains and seeds (ca. 20–40 g) were taken to determine their dry mass at 105 °C for 24 h. The yield (t ha^{-1}) was then calculated at 15% seed moisture. The thousand-grain mass of the oats and seeds of vetch were also determined.

The spatial arrangement of the experiment, with genotypes (cultivars) and farming systems including replications, is in Supplementary Figure S1. A flowchart of the methods is in Supplementary Figure S2.

2.5. Statistical Analysis of Results

The normality of the distribution of the observed traits was tested with Shapiro–Wilk's normality test to check whether the analysis of variance (ANOVA) met the assumption that the ANOVA model's residuals follow a normal distribution. Next, the effects of the main factors of the experiment: (i) farming system, (ii) oat cultivars, and (iii) years, and all the interactions between them, were estimated with a linear model for three-way ANOVA. The relationships between the traits were assessed based on Pearson's correlation coefficients and tested with the t-test. Tukey's test at $p \leq 0.05$ tested the significance of mean differences.

The results were also analyzed with multivariate methods. The canonical variate analysis (CVA) was applied to present a multi-trait assessment of the similarity of the investigated treatments in a lower number of dimensions with the least possible loss of information. This enabled the graphic illustration of the variation in the traits of all treatments under analysis. The Mahalanobis distance was suggested as a measure of similarity of multi-trait treatments, whose significance was verified employing critical value D_{cr} known as the least significant distance. Pearson's simple correlation coefficients were estimated between values of the first two canonical variates and values of the original individual traits to determine the relative share of each original trait in the multivariate variation of the treatments [34]. The GenStat v. 18 statistical software package was used for all the analyses. The GenStat v. 18 codes that have been implemented for the analyses are in Appendix A.

2.6. Weather Conditions

The weather data were collected from a meteorological station located in the Experimental Station Mydlniki-Kraków, Poland.

The sums of precipitation and the average daily air temperature in 2012–2014 differed from the standard multiyear period (1951–2000).

The humidity conditions (Figure 2) are based on the monthly precipitation for each study year. The distribution of precipitation in individual months is important for grain-legume mixture development. According to [35], the total rainfall during the vegetation period of oats in a temperate climate should range from 270 to 400 mm. The water demands of oats increase during their growth, reaching their highest values in June and July. The common vetch also has a high water demand, especially during flowering.

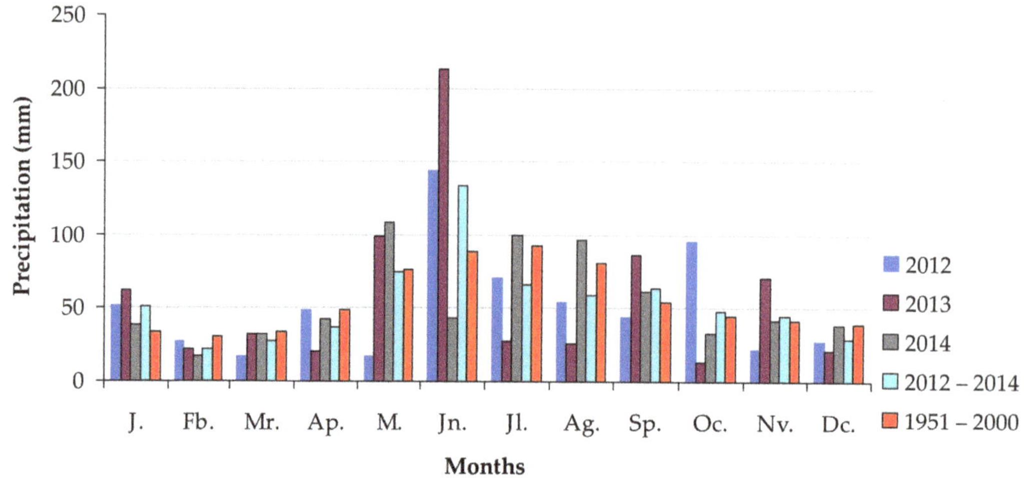

Figure 2. Sum of precipitation (mm) during the study.

The amount of precipitation in individual months and years was characterized according to the criterion of [36] for southern Poland, which classifies each month and year as "regular", or as one of three levels of "dryness", or as one of three levels of "excessive rainfall". The April–August of 2012 were dry (86% of the norm). During this year, the months of April and July were regular, May was very dry, August was dry, and June was very humid. The April–August of 2013 were classified as regular (99% of the norm). However, during this year, a large variation in precipitation was found, e.g., the months of April, July, and August were defined as very dry, May was humid, and June was extremely humid. The April–August of 2014 was regular (100.1% of the norm), with May classified as "wet" and June as "very dry" (Figure 3).

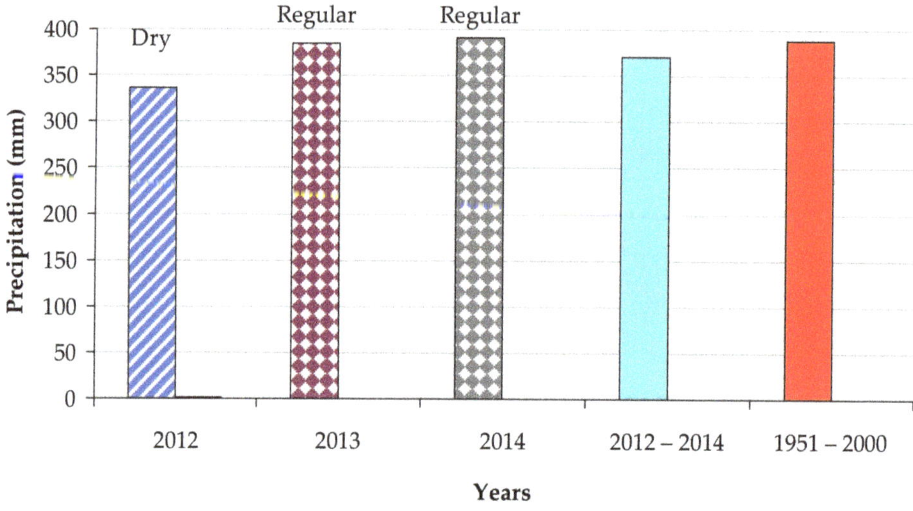

Figure 3. The sum of precipitation (mm) in the April–August period in the years of study 2012–2014 compared to the multiyear (1951–2000). Descriptors dry and regular correspond to April–August periods of the 1951–2000 multiyear.

Large fluctuations in the air temperature were observed in individual months and years of the study (Figure 4). In all study years, the average temperature (°C) was higher than the standard multiyear period (1951–2000). The air temperature was classified based on deviations in individual months of the April–August period from the norm for Krakow (Poland), according to [37]. April and June 2012 were warm, while May, July, and August were very warm. In 2013, April and August were regular, May and June were warm, and July was extremely warm. In 2014, April was warm, and May, June, and August were regular. July 2014 was an extremely warm month.

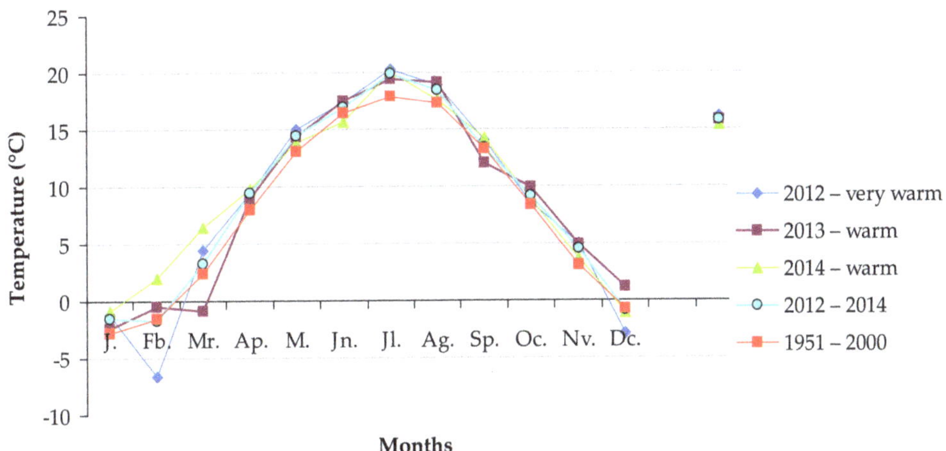

Figure 4. Mean temperatures (°C) during the study and in the 1951–2000 multiyear. Descriptors very warm and warm correspond to the 1951–2000 multiyear.

3. Results

In our study, all quantitative traits had a normal distribution. The ANOVA indicated a statistically significant influence of years and the years' × cultivar interaction for all eleven traits (Table S1).

3.1. Leaf Area and Leaf Greenness Indices

The leaf area index (LAI_1) of the oat-vetch mixture measured in the tillering phase of oats was significantly differentiated (Table 1). The LAI_1 of the mixtures in the conventional farming system was significantly higher (by 60%) than in the organic one. Additionally, the LAI_1 was affected by the weather conditions, being the highest in the optimal year 2014 (1.60 m^2 m^{-2}), and the lowest in the year 2013 (0.90 m^2 m^{-2}), most probably due to a very dry April (Figure 4).

Interactions also differentiated the LAI_1. Particularly, the interaction of oat cultivars and years was important, i.e., a significantly larger LAI_1 was found in the mixture with cv. Celer in 2012, cv. Grajcar in 2013, and in 2014 the LAI_1 was similar for both mixtures.

The LAI_2 of the oat and vetch mixtures, measured at oats' grain watery ripe (BBCH 71), was also significantly differentiated by the examined factors (Table 1). A higher LAI_2 was again found in the conventional farming; however, the system's difference diminished to 5%. Additionally, on average, the LAI_2 of the mixture with oats cv. Celer was 6% higher, compared to the one with cv. Grajcar. It is worth mentioning that the LAI_2 of mixtures with cv. Celer was similar, regardless of the farming system, whereas the LAI_1 and LAI_2 of mixtures with cv. Grajcar were higher in the conventional system by 41 and 11% compared to the organic one. The highest LAI_2 value was again in a regular year, 2014, and the lowest in a dry 2012 year.

Table 1. Leaf area index (m^2 m^{-2}) of the oat-vetch mixture, measured at the oats tillering: LAI_1 and the grain watery ripe phase; LAI_2 for the farming system in 2012–2014.

Farming System	Years	LAI_1 at the Tillering of Oats			LAI_2 at the Oats Grain Watery Ripe Phase		
		Oat Cultivar		Mean $\pm$ SD [1]	Oat Cultivar		Mean $\pm$ SD
		Celer	Grajcar		Celer	Grajcar	
Organic	2012	1.03	0.89	0.96 $\pm$ 0.10	1.94	0.91	1.43 $\pm$ 0.73
	2013	0.34	1.11	0.73 $\pm$ 0.54	1.81	1.73	1.77 $\pm$ 0.05
	2014	1.19	1.09	1.14 $\pm$ 0.07	4.33	4.49	4.41 $\pm$ 0.11
	Mean $\pm$ SD	0.86 $\pm$ 0.45	1.03 $\pm$ 0.12	0.94 B	2.69 $\pm$ 1.42	2.38 $\pm$ 1.88	2.53 B
Conventional	2012	1.66	1.09	1.37 $\pm$ 0.40	1.24	0.99	1.12 $\pm$ 0.17
	2013	0.84	1.30	1.07 $\pm$ 0.32	1.45	2.04	1.75 $\pm$ 0.42
	2014	2.18	1.95	2.07 $\pm$ 0.16	5.28	4.91	5.09 $\pm$ 0.26
	Mean $\pm$ SD	1.56 $\pm$ 0.67	1.45 $\pm$ 0.45	1.50 A	2.65 $\pm$ 2.27	2.65 $\pm$ 2.03	2.65 A
Mean	2012	1.34	0.99	1.17 $\pm$ 0.25 y	1.59	0.95	1.27 $\pm$ 0.45 z
	2013	0.59	1.21	0.90 $\pm$ 0.43 z	1.63	1.89	1.76 $\pm$ 0.18 y
	2014	1.69	1.52	1.60 $\pm$ 0.12 x	4.80	4.70	4.75 $\pm$ 0.07 x
	Mean $\pm$ SD	1.21 $\pm$ 0.56	1.24 $\pm$ 0.27	1.22	2.67 $\pm$ 1.84 a	2.51 $\pm$ 1.95 b	2.59
LSD $_{0.05}$ system		0.108			0.104		
LSD $_{0.05}$ cultivar		ns [2]			0.101		
LSD $_{0.05}$ years		0.133			0.128		
LSD $_{0.05}$ system $\times$ cultivar		0.117			0.142		
LSD $_{0.05}$ system $\times$ years		0.187			0.180		
LSD $_{0.05}$ cultivar $\times$ year		0.175			0.179		

[1] SD: standard deviation; [2] ns: non-significant. Homogeneous groups were created for the main factors. Mean values marked with the same letters are not significantly different according to Tukey's test at a significance level $p \leq 0.05$; Three-factors of experiment: (1) farming system variant—organic or conventional (letters A, B); (2) oat cultivars—Celer or Grajcar (letters a, b); (3) years—2012, 2013, and 2014 (letters x–z).

The oats' leaf relative chlorophyll content (SPAD) was differentiated by the examined factors and their interactions (Table 2). In the oats tillering phase (o_1), the oats leaf greenness index in the organic farming system was 6% higher than in the conventional farming. However, in the second term (o_2), the difference between the farming systems diminished. Additionally, a significant difference was noted between the oats' cultivars. Each time, higher SPAD values were found for the oats cv. Celer as compared to cv. Grajcar.

Table 2. The leaf chlorophyll content (relative content of chlorophyll) of oats in the mixtures with vetch, $SPADo_1$—measured at oats tillering and $SPADo_2$—measured at oats grain watery ripe phase, depending on the farming system and the oat cultivar in 2012–2014.

Farming System	Years	$SPADo_1$ at the Tillering of Oats			$SPADo_2$ at the Oats Grain Watery Ripe Phase		
		Cultivar of Oats		Mean $\pm$ SD	Cultivar of Oats		Mean $\pm$ SD
		Celer	Grajcar		Celer	Grajcar	
Organic	2012	39.8	38.2	39.0 $\pm$ 1.12	63.9	63.0	63.4 $\pm$ 0.62
	2013	40.0	35.9	37.9 $\pm$ 2.93	29.9	36.8	33.4 $\pm$ 4.85
	2014	43.4	44.8	44.1 $\pm$ 0.95	48.0	51.2	49.6 $\pm$ 2.20
	Mean $\pm$ SD	41.1 $\pm$ 2.07	39.6 $\pm$ 4.63	40.3 A	47.3 $\pm$ 17.0	50.3 $\pm$ 13.1	48.8
Conventional	2012	30.3	39.7	35.0 $\pm$ 6.68	63.1	43.2	53.1 $\pm$ 14.1
	2013	41.7	35.1	38.4 $\pm$ 4.64	44.4	37.4	40.9 $\pm$ 4.95
	2014	42.5	38.6	40.6 $\pm$ 2.77	51.8	47.3	49.6 $\pm$ 3.12
	Mean $\pm$ SD	38.1 $\pm$ 6.85	37.8 $\pm$ 2.42	38.0 B	53.1 $\pm$ 9.41	42.6 $\pm$ 4.98	47.9
Mean	2012	35.0	38.9	37.0 $\pm$ 2.78 y	63.5	53.1	58.3 $\pm$ 7.36 x
	2013	40.8	35.5	38.1 $\pm$ 3.79 y	37.2	37.1	37.1 $\pm$ 0.05 z
	2014	43.0	41.7	42.3 $\pm$ 0.91 x	49.9	49.2	49.6 $\pm$ 0.46 y
	Mean $\pm$ SD	39.6 $\pm$ 4.13 a	38.7 $\pm$ 3.13 b	39.2	50.2 $\pm$ 13.2 a	46.5 $\pm$ 8.33 b	48.3
LSD $_{0.05}$ system		1.59			ns		
LSD $_{0.05}$ cultivar		0.869			2.21		
LSD $_{0.05}$ years		1.41			3.05		
LSD $_{0.05}$ system $\times$ cultivar		ns			2.95		
LSD $_{0.05}$ system $\times$ years		1.99			4.03		
LSD $_{0.05}$ cultivar $\times$ year		1.84			4.16		

For explanation, see Table 1.

An interesting pattern was found for the oats' SPAD concerning the years. In the oats' tillering phase, higher chlorophyll content was noted in a regular 2014 year; however, in the watery ripe phase, the oats' SPAD values were highest in the dry and warm 2012, i.e., by 18% compared to the 2014 year.

The chlorophyll content of the vetch was also significantly differentiated (Table 3). Contrary to oats, higher SPAD values for vetch were found in the conventional system, compared to the organic one, by 4% in v_1 and v_2 terms. A selection of oat cultivars to the mixture with vetch also differentiated the vetch's chlorophyll content; in the v_1 term, it was higher in the mixture with cv. Celer in comparison to the v_2 term in the mixture with cv. Grajcar.

Table 3. The leaf chlorophyll content (relative content of chlorophyll) of vetch in the mixtures with oats measured at oats tillering (SPADv$_1$) and oats grain watery ripe phase (SPADv$_2$), depending on the farming system and oats cultivar in 2012–2014.

Farming System	Years	SPADv$_1$ at the Tillering of Oats			SPADv$_2$ at the Oats Grain Watery Ripe Phase		
		Cultivar of Oats		Mean ± SD	Cultivar of Oats		Mean ± SD
		Celer	Grajcar		Celer	Grajcar	
Organic	2012	39.6	39.6	39.6 ± 0.01	37.0	36.4	36.7 ± 0.37
	2013	35.5	34.7	35.1 ± 0.53	45.7	41.0	43.4 ± 3.32
	2014	38.9	39.2	39.0 ± 0.22	45.4	46.6	46.0 ± 0.83
	Mean ± SD	38.0 ± 2.20	37.8 ± 2.69	37.9 B	42.7 ± 4.97	41.3 ± 5.10	42.0 B
Conventional	2012	44.6	33.0	38.8 ± 8.20	39.1	39.1	39.1 ± 0.00
	2013	42.1	39.1	40.6 ± 2.15	39.0	51.3	45.2 ± 8.66
	2014	40.6	37.4	39.0 ± 2.28	45.6	47.4	46.5 ± 1.27
	Mean ± SD	42.4 ± 1.99	36.5 ± 3.14	39.5 A	41.3 ± 3.77	45.9 ± 6.21	43.6 A
Mean	2012	42.1	36.3	39.2 ± 4.11	38.0	37.8	37.9 ± 0.19 z
	2013	38.8	36.9	37.8 ± 1.34	42.4	46.1	44.3 ± 2.67 y
	2014	39.8	38.3	39.0 ± 1.03	45.5	47.0	46.3 ± 1.05 x
	Mean ± SD	40.2 ± 1.70 a	37.2 ± 1.04 b	38.7	42.0 ± 3.75 b	43.6 ± 5.10 a	42.8
LSD $_{0.05}$ system		1.35			0.84		
LSD $_{0.05}$ cultivar		1.50			1.34		
LSD $_{0.05}$ years		ns			1.02		
LSD $_{0.05}$ system × cultivar		2.03			1.57		
LSD $_{0.05}$ system × years		2.26			ns		
LSD $_{0.05}$ cultivar × year		2.26			1.44		

For explanation, see Table 1.

An interesting pattern of vetch's chlorophyll content was noted concerning the years. In the v_1 term, the SPAD of the vetch was similar for all the years. Contrarily, in the v_2 term, the highest vetch SPAD values were noted in a regular year, 2014, and the lowest were noted in the dry 2012. This is the reverse of the oat's SPAD values in the same term (SPADo$_2$) (Table 4).

Table 4. Seed yield (t ha^{-1}) of oat-vetch mixtures depending on the farming system and oat cultivar in 2012–2014.

Farming System	Years	Cultivar of Oats		Mean ± SD
		Celer	Grajcar	
Organic	2012	4.10	4.13	4.12 ± 0.02
	2013	2.17	2.08	2.13 ± 0.06
	2014	2.31	2.44	2.38 ± 0.09
	Mean ± SD	2.86 ± 1.08	2.88 ± 1.09	2.87 B
Conventional	2012	4.57	4.13	4.35 ± 0.31
	2013	2.65	3.21	2.93 ± 0.40
	2014	4.24	3.88	4.06 ± 0.25
	Mean ± SD	3.82 ± 1.02	3.74 ± 0.47	3.78 A
Mean	2012	4.34	4.13	4.23 ± 0.15 x
	2013	2.41	2.65	2.53 ± 0.17 z
	2014	3.28	3.16	3.22 ± 0.08 y
	Mean ± SD	3.34 ± 0.96	3.31 ± 0.75	3.33
LSD $_{0.05}$ system			0.038	
LSD $_{0.05}$ cultivar			ns	
LSD $_{0.05}$ years			0.106	
LSD $_{0.05}$ system × cultivar			ns	
LSD $_{0.05}$ system × years			0.128	
LSD $_{0.05}$ cultivar × year			0.137	

For explanation, see Table 1.

3.2. Yield of Mixtures and Their Components

The mixtures' yield was 24% higher in the conventional system than the organic one (Table 4). An interaction was found for oat cultivars and years, e.g., the yield of the mixture with oats cv. Celer was significantly higher in a dry 2012 and a regular 2014, compared to 2013.

A significantly higher, by 38%, share of vetch seeds in the seed yield of mixtures was found in the organic system compared to the conventional one (Table 5). Additionally, on average, a higher share of vetch seeds was found in the mixture with oats cv. Grajcar, compared to oats cv. Celer. The share of vetch seeds in the yield was lowest in the dry 2012 and highest in the year 2013.

Table 5. The share of vetch seeds (%) in the oat-vetch mixture yields depending on the farming system and oat cultivar in 2012–2014.

Farming System	Years	Cultivar of Oats		Mean ± SD
		Celer	Grajcar	
Organic	2012	5.18	23.3	14.2 ± 12.8
	2013	76.3	70.7	73.5 ± 3.97
	2014	60.6	65.3	63.0 ± 3.28
	Mean ± SD	47.4 ± 37.4	53.1 ± 25.9	50.2 A
Conventional	2012	1.21	3.91	2.56 ± 1.91
	2013	51.8	45.0	48.4 ± 4.80
	2014	29.7	54.4	42.0 ± 17.4
	Mean ± SD	27.6 ± 25.3	34.4 ± 26.8	31.0 B
Mean	2012	3.20	13.6	8.40 ± 7.36 z
	2013	64.0	57.8	60.9 ± 4.38 x
	2014	45.2	59.8	52.5 ± 10.4 y
	Mean ± SD	37.5 ± 31.1 b	43.7 ± 26.1 a	40.6
LSD $_{0.05}$ system			2.70	
LSD $_{0.05}$ cultivar			2.85	
LSD $_{0.05}$ years			2.88	
LSD $_{0.05}$ system × cultivar			ns	
LSD $_{0.05}$ system × years			4.07	
LSD $_{0.05}$ cultivar × year			4.07	

For explanation, see Table 1.

Oats produced more tillers per plant and more panicles per unit area in the conventional system (Table 6). Interestingly, even though oats cv. Grajcar produced more tillers in the mixture, as compared to the oats cv. Celer, Grajcar still had a lower number of panicles per area in comparison with Celer. The highest number of oats' tillers and panicles was noted for both cultivars and farming systems in the dry year 2012. Despite a similar number of oats' tillers in 2013 and 2014, there was a significant drop in the number of oat panicles per unit area in 2013, regardless of the farming system and oat cultivar.

Like the seed yield and the number of panicles per area, a significantly greater number of grains per oat panicle (by 31%) were present in the conventional system compared to the organic one (Table 7)—oats cv. Celer developed by 38% more grains per panicle in the mixtures, compared to the cv. Grajcar. It was found that the number of grains of cv. Celer was significantly higher in conventional farming, by 43%, compared to the organic one, whereas the number of grains of the cv. Grajcar was similar in both farming systems. The number of grains in the panicles was highest in the regular year 2014. In the other two years, the number of grains per panicle was similar.

Table 6. The average number of tillers per oat plant and number of oats panicles per m^{-2} in the oat-vetch mixtures, depending on the farming system and oat cultivar in 2012–2014.

| Farming System | Years | The Average Number of Oats' Tillers | | | The Number of Oats Panicles per m^{-2} | | |
| | | Cultivar of Oats | | Mean ± SD | Cultivar of Oats | | Mean ± SD |
		Celer	Grajcar		Celer	Grajcar	
Organic	2012	1.37	1.70	1.54 ± 0.23	602	431	517 ± 121
	2013	1.24	1.40	1.32 ± 0.11	108	156	132 ± 33.9
	2014	1.28	1.35	1.31 ± 0.05	208	208	208 ± 0.00
	Mean ± SD	1.29 ± 0.07	1.48 ± 0.19	1.39 B	306 ± 261	265 ± 146	286 B
Conventional	2012	1.73	1.83	1.78 ± 0.07	645	475	560 ± 120
	2013	1.55	1.73	1.64 ± 0.12	123	136	129 ± 9.43
	2014	1.43	1.48	1.45 ± 0.04	313	296	305 ± 12.0
	Mean ± SD	1.57 ± 0.15	1.68 ± 0.18	1.62 A	360 ± 264	302 ± 170	331 A
Mean	2012	1.55	1.76	1.66 ± 0.15 x	624	453	538 ± 121 x
	2013	1.40	1.56	1.48 ± 0.12 y	115	146	131 ± 21.7 z
	2014	1.35	1.41	1.38 ± 0.04 y	261	252	256 ± 6.01 y
	Mean ± SD	1.43 ± 0.10 b	1.58 ± 0.18 a	1.51	333 ± 262 a	284 ± 156 b	308
LSD $_{0.05}$ system		0.159			10.9		
LSD $_{0.05}$ cultivar		0.109			8.97		
LSD $_{0.05}$ years		0.101			15.1		
LSD $_{0.05}$ system × cultivar		ns			ns		
LSD $_{0.05}$ system × years		ns			20.4		
LSD $_{0.05}$ cultivar × year		ns			19.6		

For explanation, see Table 1.

Table 7. The number of grains per oat panicle in the oat-vetch mixtures, depending on the farming system and oat cultivar in 2012–2014.

| Farming System | Years | Cultivar of Oats | | Mean ± SD |
		Celer	Grajcar	
Organic	2012	13.2	20.8	17.0 ± 5.37
	2013	21.6	7.7	14.7 ± 9.80
	2014	19.2	15.8	17.5 ± 2.38
	Mean ± SD	18.0 ± 4.33	14.8 ± 6.60	16.4 B
Conventional	2012	16.8	16.7	16.7 ± 0.04
	2013	30.6	14.3	22.4 ± 11.5
	2014	47.9	17.0	32.5 ± 21.8
	Mean ± SD	31.7 ± 15.6	16.0 ± 1.50	23.9 A
Mean	2012	15.0	18.7	16.9 ± 2.67 y
	2013	26.1	11.0	18.5 ± 10.7 y
	2014	33.6	16.4	25.0 ± 12.1 x
	Mean ± SD	24.9 ± 9.35 a	15.4 ± 3.97 b	20.1
LSD $_{0.05}$ system		2.85		
LSD $_{0.05}$ cultivar		3.98		
LSD $_{0.05}$ years		4.40		
LSD $_{0.05}$ system × cultivar		4.86		
LSD $_{0.05}$ system × years		5.80		
LSD $_{0.05}$ cultivar × year		6.22		

For explanation, see Table 1.

The number of vetch pods per m^{-2} and the number of vetch seeds per pod (Table 8) followed, to some extent, the pattern of the share of vetch seeds in the mixture's yield (Table 5). Compared to the conventional system, the number of vetch pods was 53% higher in the organic one. The highest number of vetch pods was found in 2013 in the mixture with cv. Grajcar. However, a significantly higher number of seeds per pod was noted in conventional farming over organic. The highest number of vetch seeds per pod was found in the mixture with cv. Grajcar in the regular year 2014. The weather also influenced the vetch pod and seed per pod production in a significant way. Interestingly, the highest number of pods per m^{-2} was found in the 2013 year, but the highest number of seeds per pod was found in the regular 2014 year (Table 8).

Table 8. The pod number per m^2 and seed number per pod of vetch grown in the oat-vetch mixtures, depending on the farming system and oat cultivar in 2012–2014.

Farming System	Years	No. of Vetch Pods per m^2				No. of Seeds per Pod		
		Cultivar of Oats		Mean ± SD		Cultivar of Oats		Mean ± SD
		Celer	Grajcar			Celer	Grajcar	
Organic	2012	152	190	171 ± 27		2.73	3.23	2.98 ± 0.35
	2013	567	948	758 ± 269		4.72	5.01	4.86 ± 0.21
	2014	414	714	564 ± 212		6.18	6.99	6.59 ± 0.57
	Mean ± SD	378 ± 210	617 ± 388	497 A		4.54 ± 1.73	5.08 ± 1.88	4.81 B
Conventional	2012	40	76	58 ± 25		2.97	2.23	2.60 ± 0.52
	2013	326	224	275 ± 72		5.01	5.18	5.10 ± 0.12
	2014	257	488	373 ± 163		7.33	7.58	7.46 ± 0.18
	Mean ± SD	208 ± 149	263 ± 209	235 B		5.11 ± 2.18	5.00 ± 2.68	5.05 A
Mean	2012	96	133	115 ± 26 z		2.85	2.73	2.79 ± 0.09 z
	2013	447	586	516 ± 99 x		4.86	5.09	4.98 ± 0.16 y
	2014	335	601	468 ± 188 y		6.76	7.29	7.02 ± 0.38 x
	Mean ± SD	293 ± 179 b	440 ± 266 a	366 ±		4.82 ± 2.0 b	5.04 ± 2.28 a	4.93
LSD $_{0.05}$ system		49.6				0.07		
LSD $_{0.05}$ cultivar		31.4				0.21		
LSD $_{0.05}$ years		38.7				0.24		
LSD $_{0.05}$ system × cultivar		44.4				0.22		
LSD $_{0.05}$ system × years		54.8				0.29		
LSD $_{0.05}$ cultivar × year		54.6				0.35		

For explanation, see Table 1.

The thousand-grain mass (TGM) of oats was higher in the conventional system, whereas for vetch this was in the organic one (Table 9). Simultaneously, higher TGMs of both oats and vetch were noted in the mixtures with cv. Celer. The TGM of oat cv. Celer was similar, regardless of the farming system, but in the case of cv. Grajcar was by 7% higher in the conventional system than in the organic one. The TGM of vetch fitted well to this pattern, as it was similar in the mixture with Celer, but 13% lower in the mixture with cv. Grajcar in the conventional system, compared to the organic one. On average, the TGM of both oats and vetch was lowest in the dry 2012 and highest in the regular 2014 year.

Table 9. The thousand-grain mass (TGM) of oats and vetch (g) in the oat-vetch mixture, depending on the farming system and oat cultivar in 2012–2014.

Farming System	Years	TGM of Oats				TGM of Vetch		
		Cultivar of Oats		Mean ± SD		Cultivar of Oats		Mean ± SD
		Celer	Grajcar			Celer	Grajcar	
Organic	2012	37.5	30.1	33.8 ± 5.22		51.6	51.9	51.8 ± 0.17
	2013	42.9	33.4	38.1 ± 6.75		54.5	56.4	55.5 ± 1.34
	2014	46.5	35.9	41.2 ± 7.51		63.4	58.8	61.1 ± 3.22
	Mean ± SD	42.3 ± 4.56	33.1 ± 2.92	37.7 B		56.5 ± 6.10	55.7 ± 3.51	56.1 A
Conventional	2012	36.2	30.4	33.3 ± 4.08		49.9	40.1	45.0 ± 6.94
	2013	43.4	33.8	38.6 ± 6.78		50.7	54.6	52.7 ± 2.71
	2014	47.7	42.0	44.8 ± 3.98		59.6	53.2	56.4 ± 4.53
	Mean ± SD	42.4 ± 5.81	35.4 ± 5.98	38.9 A		53.4 ± 5.36	49.3 ± 8.00	51.3 B
Mean	2012	36.8	30.2	33.5 ± 4.65 z		50.8	46.0	48.4 ± 3.38 z
	2013	43.2	33.6	38.4 ± 6.76 y		52.6	55.5	54.1 ± 2.03 y
	2014	47.1	39.0	43.0 ± 5.75 x		61.5	56.0	58.7 ± 3.87 x
	Mean ± SD	42.4 ± 5.18 a	34.3 ± 4.40 b	38.3		54.9 ± 5.71 a	52.5 ± 5.64 b	53.7
LSD $_{0.05}$ system		1.00				0.614		
LSD $_{0.05}$ cultivar		0.572				1.06		
LSD $_{0.05}$ years		1.43				0.857		
LSD $_{0.05}$ system × cultivar		0.809				1.22		
LSD $_{0.05}$ system × years		1.92				1.21		
LSD $_{0.05}$ cultivar × year		1.74				1.21		

For explanation, see Table 1.

The Pearson correlation coefficient analyses revealed several statistically significant interdependencies between the observed traits (Table S2, Figure 5). LAI_1 (leaf area index in the oats' tillering phase BBCH 29) was significantly positively correlated with: LAI_2, leaf area index in the oats BBCH 71 phase; $SPADo_2$, relative chlorophyll content in oat leaves in the oats BBCH 71 phase; yd, mixtures yield; no-p, number of oats panicles per m^2; no-gr, number of oats grains per panicle, and no-sd, number of vetch seeds per pod. LAI_2

was positively correlated with: SPADv$_2$, relative chlorophyll content in vetch leaves in the oat BBCH 71 phase; sh-v, share of vetch in the mixture's yield; no-gr; TGWo, thousand-grain mass of oats; TGWv, thousand-grain mass of vetch; no-pod, number of vetch pods per m^2; and no-sd. SPADo$_2$ was positively correlated with: yd and no-p; and negatively correlated with: SPADv$_2$, sh-v, and no-pod. SPADv$_2$ was positively correlated with: sh-v, TGWo, TGWv, no-pod, and no-sd; and negatively with yd and no-p. The yd was positively correlated with no-p and negatively correlated with sh-v, TGWv, no-pod, and no-sd. The sh-v was negatively correlated with no-p (-0.691) and positively with TGWo, TGWv, no-pod, and no-sd. The no-p negatively correlated with no-pod and no-sd. TGWo was positively correlated with no-gr, TGWv, and no-sd. TGWv positively correlated with no-pod and no-sd, and additionally, no-sd correlated with no-pod. SPADo$_1$ was positively correlated with: LAI$_2$, sh-v, no-gr, TGWo, TGWv, and no-sd; and negatively with yd and no-p. SPADv$_1$ correlated positively with SPADo$_2$ and SPADv$_1$; and negatively with no-pod (Figure 5, Table S2).

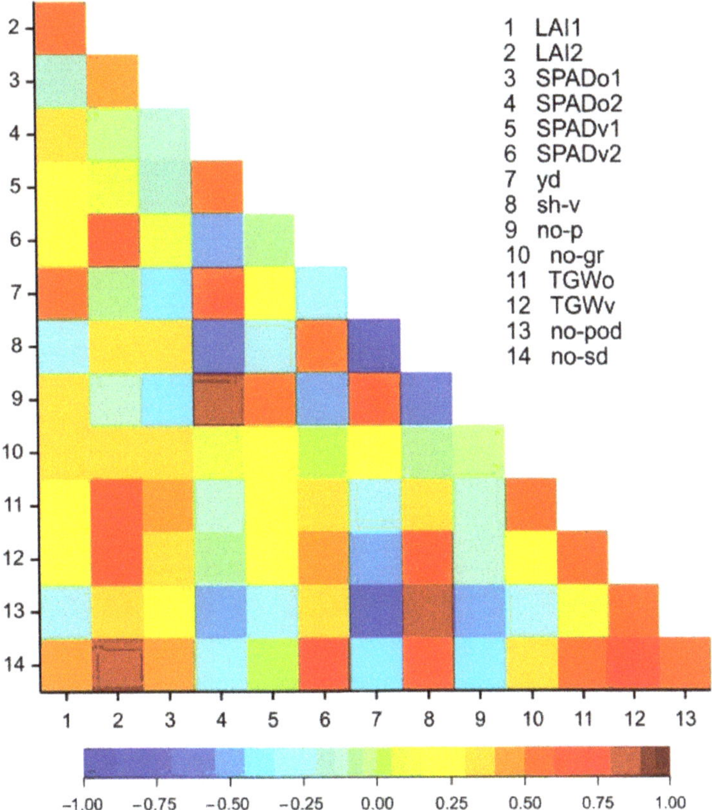

Figure 5. Heatmap for linear Pearson's correlation coefficients between observed traits; $r_{cr} = 0.2875$.

The greatest diversity in all eleven traits, measured with Mahalanobis distances, was observed for the combination co-ce-12 (conventional variant-Celer-2012) and or-ce-13 (organic variant-Celer-2013) (Table S3). The Mahalanobis distance between them amounted to 74.44. The greatest similarity (distance: 11.73) was observed between co-ce-14 (conventional variant-Celer-2014) and co-gr-14 (conventional variant-Grajcar-2014).

The canonical analysis was performed to present the tested mixtures' overall performances, based on all of the tested traits, for all of the three factors of this experiment

(Figure 5). The first two canonical variates explained jointly 85.6% of the total variation between the treatments. The greatest, significant linear relationship with the first canonical variate was found for SPADv$_2$, the share of vetch in the mixture's yield, TGWv, number of vetch pods, number of vetch seeds per pod (positive dependencies), and SPADo2, the yield of mixture and number of panicles per m^2 (negative dependencies). The second canonical variate was significantly positively correlated with LAI$_1$, LAI$_2$, and the number of vetch seeds per pod. The results point to the best performance of the mixtures in the conventional variant of the farming system and during the regular year 2014 (Figure 6). However, both mixtures performed well also in the organic system in 2014. The mixtures performed worst in both organic and conventional systems in 2012.

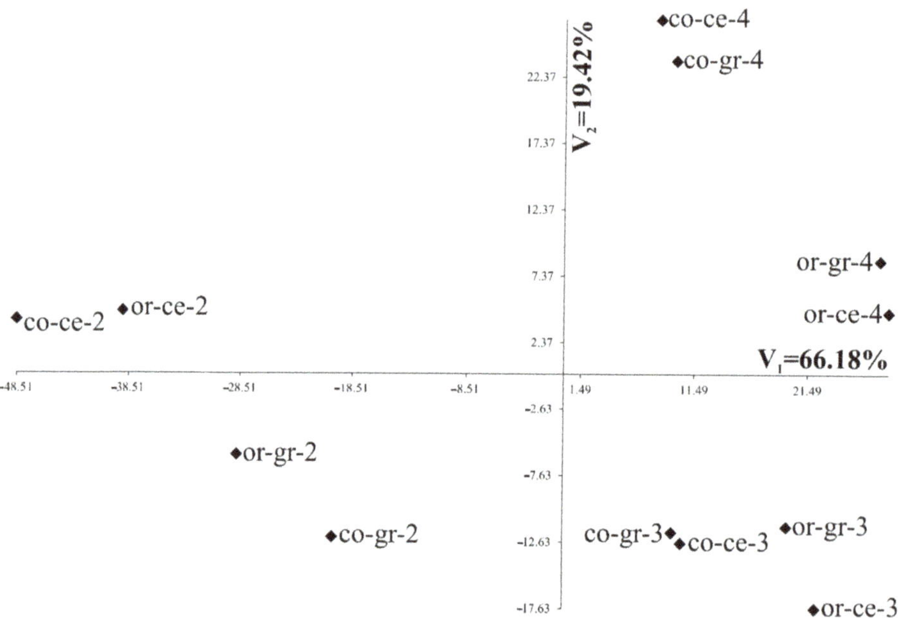

Figure 6. The distribution of all 12 combinations of farming systems, cultivars, and years of study in the two first canonical variates, based on all tested traits. In the diagram, the coordinates of a given combination of treatments are values of the first and second canonical variates. Co: conventional farming; or: organic farming; ce: Celer, gr: Grajcar; 2–4: years 2012–2014.

4. Discussion

The farming system affected the seed yield of mixtures by approximately 24% in favor of the mixtures grown conventionally, compared to those grown organically. These findings are consistent with several other studies [4,37–43] and result mainly from the direct growth- and yield-promoting effects of mineral nutrition of crops in the conventional system. However, Schram et al. [44] underline that crop yield differences between farming systems diminish with time; after 13 years, they amount to only 13% in favor of the conventional system over the organic one.

A detailed analysis revealed that the mixtures' components, namely oats and common vetch, reacted differently to agricultural production intensification. The share of vetch seeds in the seed yield, number of pods per m^2, and the thousand-seed mass of vetch were higher in an organic farming system. Reversely, oats yielded well in the conventional system. Under stressful conditions of a limited supply of soil resources, the legume component performs better than the cereal one, leading to the resilience of a total mixture yield [45]. Due to an extensive root system, legumes can activate phosphorus from organic compounds in the soil, mostly unavailable to cereals [46]. Moreover, they also use biologically bound nitrogen assimilated by the *Rhizobium* bacteria [47]. This effect clearly shows a complementarity of the components of the oat-vetch mixture. A proper selection of cereal components for mixture with a legume is of significance in this context. The interaction of oat cultivars and farming system variant, and the oat cultivar and year were observed in our study for almost all of the analyzed plant and canopy traits. In general, oats cv. Celer turned to be more competitive toward vetch in the mixture as compared to oats cv. Grajcar. Interestingly, both oat cultivars tested in this study were characterized by their breeder as having a very similar set of traits, i.e., time to ripening, thousand-grain mass, and plant height. The competitive effort of oat cultivars toward vetch was related to their productivity traits—specifically, even though oats cv. Grajcar developed more tillers in the mixture, as compared to cv. Celer they were less productive, i.e., displaying a lower density of panicles per m^{-2}, a lower number of grains per panicle, and a lower thousand-grain mass. Contrarily, vetch was more productive in the mixture with cv. Grajcar as reflected by a higher number of vetch pods per m^2, seeds per pod, and a share of vetch seeds in the mixture's yield, compared to the mixture with cv. Celer. As a result, even though both mixtures had a similar total yield during the years of study, the mixture of oats cv. Grajcar and vetch cv. Hanka had a more optimal share of oats/vetch seeds in the yield than the mixture with cv. Celer. Noteworthy was the finding of the negative correlation of the mixture yield with the number of pods and percentage of vetch seeds in the mixture yield. The greater the yield of the mixture, the lower the percentage of vetch seeds. Contrarily, the lower the mixture's yield, the greater the number of vetch pods per unit area. Both findings indicate strong competitive effects of oats toward vetch. Only a few studies discuss the influence of oat cultivar selection on the yield of the oat-vetch mixture, e.g., [48]. In our previous studies, we have shown that the oat cultivar is crucial for a good vetch yield, which is also influenced by the type of soil [49]. The share of vetch seeds in the mixture with oats is variable and influenced by several factors [50–52]. The main restrictions are weather conditions during the growing year. With low rainfall, vetch cannot withstand competition for water with oats, and its share in the yield is smaller [51–53].

In general, the leaf area index, which relates to the leaf assimilation area, and the leaf relative chlorophyll content (SPAD) were higher for the mixtures grown in the conventional system compared to the organic one. The LAI and chlorophyll content measured in SPAD values are good indices of the crop canopy status; many authors confirm their usefulness for estimating crop yields [54,55].

The results of the canonical analysis, performed for all of the tested factors, revealed that weather conditions were the main driver affecting the performance of the mixtures. The best year turned out a regular year, namely 2014. In 2012, a severe drought occurred in May and later in July, whereas June was very humid. In that year, regardless of oat cultivar and farming system variant, oats over-compete vetch by developing a significantly higher number of panicles than in 2013 and 2014. This shows that both oat cultivars tend to redistribute assimilates to produce higher grain yields in stressful conditions. This finding agrees with Zao et al. [56], who found a similar phenomenon in oats cv. Bia. According to those authors, under moderate drought stress there is a decreased biomass distribution to stems and leaves and a greater grain yield of oats. On the other hand, in 2013, when an excess of precipitation occurred in May and June and a severe drought in July and August, the share of vetch seeds in the mixtures' yields was the highest and for oats this yield was the lowest. These results confirm the benefits of cultivating mixtures, namely maintaining a high yield of at least one mixture component in years with weather unfavorable for the other component of the mixture [57].

5. Conclusions

A greater share (by 62%) of vetch seeds in the mixture yield and a greater thousand-seeds mass of vetch (by 9.3%) was noted in the organic system. The proper selection of oat cultivar for mixing with vetch may support a higher share of vetch seeds in the yield.

In this research, the less productive cultivar (with a lower number of panicles per m^{-2} and grains per panicle) was a better companion for vetch in the mixture. This study revealed that temperature and precipitation affect the final performance of the oat-vetch mixture. Under adverse weather conditions, a changeable share of both components of the mixture led to the yield compensation.

The canopy indices of the mixtures, LAI and SPAD, are diversified. However, the type of farming system and the oat cultivar selection significantly impact these traits. The LAI, SPAD, and the seed yield of mixtures were higher in the conventional farming system.

Summing up, the oat-vetch mixture is recommended for organic farming. However, the proper selection of the cereal component for this mixture is of high importance.

Supplementary Materials: The following are available online at https://www.mdpi.com/article/10.3390/agriculture11040332/s1, Figure S1: The spatial arrangement of replication with genotypes and management systems, Figure S2: Flowchart of methodology of the research; Table S1: Mean squares from three-way analysis of variance (ANOVA) for observed traits; Table S2: Correlation coefficients between the quantitative traits; Table S3: Mahalanobis distances between pairs of combinations of the three studied factors.

Author Contributions: Conceptualization, K.P. and S.P.; methodology, K.P. and S.P.; validation, A.L., K.K.; formal analysis, K.P., S.P., K.K., A.L., A.S.; investigation, K.P. and S.P.; data curation, K.P., S.P., J.B., and A.S.; writing—original draft preparation, K.P., A.S., S.P., writing—review and editing, KP, S.P., J.B., A.S.; visualization, K.P., S.P., J.B., A.S.; funding acquisition, K.P. and A.L. All authors have read and agreed to the published version of the manuscript.

Funding: This research was financed by the Ministry of Science and Higher Education of the Republic of Poland–partially, this research was funded by the Ministry of Science and Higher Education in Poland, for education in the years 2010–2015 as research project (grant no. N N310 446938).

Institutional Review Board Statement: Not applicable.

Informed Consent Statement: Not applicable.

Data Availability Statement: The data presented in this study are available on request from the corresponding authors.

Conflicts of Interest: The authors declare no conflict of interest.

Appendix A

The GenStat v. 18 codes:

```
JOB
IMPORT 'Data.xls';is=names
PRINT names

FOR Trait=LAI1,LAI2,SPADo1,SPADo2,SPADv1,SPADv2,yd,sh_v,no_p,no_gr,TGWo,
TGWv,no_pod,no_sd
WSTATISTIC [p=test] Trait
ENDFOR

FOR Trait=LAI1,LAI2,SPADo1,SPADo2,SPADv1,SPADv2,yd,sh_v,no_p,no_gr,TGWo,
TGWv,no_pod,no_sd

TREAT Year*System*Cultivar

BLOCK Repl/DuPol/MaPol

ANOVA [p=aovt,mean;fprob=y;pse=lsd;FACTORIAL=5] Trait
ENDFOR
```

```
TEXT [12] Name
READ Name
co_ce_12
co_gr_12
co_ce_13
co_gr_13
co_ce_14
co_gr_14
or_ce_12
or_gr_12
or_ce_13
or_gr_13
or_ce_14
or_gr_14:

SSPM [LAI1,LAI2,SPADo1,SPADo2,SPADv1,SPADv2,yd,sh_v,no_p,no_gr,TGWo,
TGWv,no_pod,no_sd;group=Number] ssp
FSSPM ssp
CVA                          [p=roots,loadings,means,residuals,distances,tests]
ssp;scores=cvm;DISTANCES=MahP
pen number=1;Labels=Name
DGRAPH [key=0] cvm$[*;2];cvm$[*;1]

TABULATE [p=mean;cl=Number] LAI1,LAI2,SPADo1,SPADo2,SPADv1,SPADv2,
yd,sh_v,no_p,no_gr,TGWo,TGWv,no_pod,no_sd

FCORRELATION [p=corr,test] LAI1,LAI2,SPADo1,SPADo2,SPADv1,SPADv2,
yd,sh_v,no_p,no_gr,TGWo,TGWv,no_pod,no_sd

ENDJOB
```

References

1. Paut, R.; Sabatier, R.; Tchamitchian, M. Modelling crop diversification and association effects in agricultural systems. *Agric. Ecosyst. Environ.* **2020**, *288*, 106711. [CrossRef]
2. Statistics Poland. *Concise Statistical Yearbook of Poland*; Zakład Wydawnictw Statystycznych: Warsaw, Poland, 2018. Available online: https://stat.gov.pl/files/gfx/portalinformacyjny/pl/defaultaktualnosci/5515/1/19/1/maly_rocznik_statystyczny_polski_2018.pdf (accessed on 20 May 2020).
3. Kaut, A.H.E.E.; Mason, H.E.; Navabi, A.; O'Donovan, J.T.; Spaner, D. Organic and conventional management of mixtures of wheat and spring cereals. *Agron. Sustain. Dev.* **2008**, *28*, 363–371. [CrossRef]
4. Klima, K.; Łabza, T. Yielding and economic efficiency of oats crops cultivated using pure and mixed sowing stands in organic and conventional farming systems. *Żywn. Nauka. Technol. Jakość.* **2010**, *3*, 141–147.
5. Szpunar-Krok, E.; Bobrecka-Jamro, D.; Tobiasz-Salach, R. Yielding of naked oats and faba bean in Pure sowing and mixtures. *Fragm. Agron.* **2009**, *26*, 145–151.
6. Jedel, P.E.; Salmon, D.F. Forage potential of spring and winter cereal mixtures in a short-year growing area. *Agron. J.* **1995**, *87*, 731–736. [CrossRef]
7. Juskiw, P.E.; Helm, J.H.; Salmon, D.F. Forage yield and quality for monocrops and mixtures of small grain cereals. *Crop. Sci.* **2000**, *40*, 138–147. [CrossRef]
8. Omokanye, A.; Lardner, H.; Lekshmi, S.; Jerey, L. Forage production, economic performance indicators and beef cattle nutritional suitability of multispecies annual crop mixtures in northwestern Alberta, Canada. *J. Appl. Anim. Res.* **2019**, *47*, 303–313. [CrossRef]
9. Leszczyńska, D. State and conditions of cultivation of grain crops mixtures in Poland. *J. Res. Appl. Agric. Eng.* **2007**, *52*, 105–108.
10. Leszczyńska, D. Actual state and conditions of cultivation of grain crops mixtures in Poland. *J. Res. Appl. Agric. Eng.* **2010**, *55*, 7–11.
11. Staniak, M.; Księżak, J.; Bojarszczuk, J. Estimation of productivity and nutritive value of pea-barley mixtures in organic farming. *J. Food Agric. Environ.* **2012**, *10*, 318–323. [CrossRef]
12. Blesh, J.; VanDusen, B.M.; Brainard, D.C. Managing ecosystem services with cover crop mixtures on organic farms. *Agron. J.* **2019**, *111*, 826–840. [CrossRef]

13. Klima, K.; Synowiec, A.; Puła, J.; Chowaniak, M.; Pużyńska, K.; Gala-Czekaj, D.; Kliszcz, A.; Galbas, P.; Jop, B.; Dąbkowska, T.; et al. Long-term productive, competitive, and economic aspects of spring cereal mixtures in integrated and organic crop rotations. *Agriculture* **2020**, *10*, 321. [CrossRef]

14. Gentsch, N.; Boy, J.; Guggenberger, G. Incorporation of diverse catch crop mixtures in crop rotation cycles increase biodiversity and nutrient availability in soils. In *Horizonte des Bodens, Proceedings of the Jahrestagung der Deutsche Bodenkundliche Gesellschaft, Göttingen, Germany, 2–6 September 2017*; DBG: Berlin, Germany, 2017.

15. Blesh, J. Functional Traits in Cover Crop Mixtures: Biological Nitrogen Fixation and Multifunctionality. *J. Appl. Ecol.* **2018**, *55*, 38–48. [CrossRef]

16. Lori, M.; Symnaczik, S.; Mäder, P.; De Deyn, G.; Gattinger, A. Organic farming enhances soil microbial abundance and activity—a meta-analysis and meta-regression. *PLoS ONE* **2017**, *12*, e0180442. [CrossRef]

17. He, H.M.; Liu, L.N.; Munir, S.; Bashir, N.H.; Wang, Y.; Yang, J.; Li, C.Y. Crop diversity and pest management in sustainable agriculture. *J. Integr. Agric.* **2019**, *18*, 1945–1952. [CrossRef]

18. Głąb, T.; Pużyńska, K.; Pużyński, S.; Palmowska, J.; Kowalik, K. Effect of organic farming on a Stagnic Luvisol soil physical quality. *Geoderma* **2016**, *282*, 16–25. [CrossRef]

19. Saleem, M.; Pervaiz, Z.H.; Contreras, J.; Lindenberger, J.H.; Hupp, B.M.; Chen, D.; Zhang, Q.; Wang, C.; Iqbal, J.; Twigg, P. Cover Crop Diversity Improves Multiple Soil Properties via Altering Root Architectural Traits. *Rhizosphere* **2020**, *16*, 100248. [CrossRef]

20. Gaudio, N.; Escobar-Gutiérrez, A.J.; Casadebaig, P.; Evers, J.B.; Gerard, F.; Louarn, G.; Colbach, N.; Munz, S.; Launay, M.; Marrous, H.; et al. Current knowledge and future research opportunities for modeling annual crop mixtures. A review. *Agron. Sustain. Dev.* **2019**, *39*, 20. [CrossRef]

21. Pisulewska, E.; Klima, K.; Witkowicz, R.T.; Borowiec, F. Grain yield, fatty acid content and composition of oatscultivar Dukat as affected by sowing techniques. *Food. Sci. Technol. Qual.* **1999**, *1*, 246–252.

22. Rudnicki, F. Biologiczne aspekty uprawy zbóż w mieszankach. In *Stan i Perspektywy Uprawy Mieszanek Zbożowych*; AR: Poznań, Poland, 1994; pp. 7–15.

23. Kotwica, K.; Rudnicki, F. Production effects of growing spring cereal and cereal-and legume mixtures on good rye complex soil. *Acta Sci. Pol. Agric.* **2004**, *3*, 149–156.

24. Rudnicki, F.; Wenda-Piesik, A.; Wasilewski, P. Sowing rate of components in pea-barley intercropping on the wheat soil complex. *Zesz. Probl. Post. Nauk Rol.* **2007**, *516*, 195–208.

25. Klima, K.; Stokłosa, A.; Pużyńska, K. Agricultural and economic circumstances of cereal cultivation under differentiated soil and climate conditions. *Zesz. Probl. Post. Nauk Rol.* **2011**, *559*, 115–121.

26. Ceglarek, F.; Rudziński, R.; Płaza, A.; Buraczyńska, D. Nutritive value of common vetch [Vicia sativa L.] grown in pure and mixed stands in the middle-east Poland. *Zesz. Probl. Post. Nauk Rol.* **2007**, *516*, 19–26.

27. Sobkowicz, P.; Tendziagolska, E. Competition and Productivity in Mixture of Oats and Wheat. *J. Agron. Crop Sci.* **2005**, *191*, 377–385. [CrossRef]

28. Artyszak, A. Dobór komponentów i skład mieszanek z udziałem jarych roślin strączkowych uprawianych na nasiona—Przegląd literatury. *Post. Nauk Rol.* **1993**, *4*, 81–87.

29. Fang, H.; Baret, F.; Plummer, S.; Schaepman-Strub, G. An Overview of Global Leaf Area Index (LAI): Methods, Products, Validation, and Applications. *Rev. Geophys.* **2019**, *57*, 739–799. [CrossRef]

30. Mendoza-Tafolla, R.O.; Juarez-Lopez, P.; Ontiveros-Capurata, R.-E.; Sandoval-Villa, M.; Alia-Tejacal, I.; Alejo-Santiago, G. Estimating Nitrogen and Chlorophyll Status of Romaine Lettuce Using SPAD and at LEAF Readings. *Not. Bot. Horti Agrobot. Cluj Napoca* **2019**, *47*. [CrossRef]

31. Srinivasan, V.; Kumar, P.; Long, S.P. Decreasing, Not Increasing, Leaf Area Will Raise Crop Yields under Global Atmospheric Change. *Glob. Change Biol.* **2017**, *23*, 1626–1635. [CrossRef]

32. Hirooka, Y.; Homma, K.; Maki, M.; Sekiguchi, K.; Shiraiwa, T.; Yoshida, K. Evaluation of the Dynamics of the Leaf Area Index (LAI) of Rice in Farmer's Fields in Vientiane Province, Lao PDR. *J. Agric. Meteorol.* **2017**, *73*, 16–21. [CrossRef]

33. IUSS Working Group WRB. *World Reference Base for Soil Resources 2014, International Soil Classification System for Naming Soils and Creating Legends for Soil Maps—Update 2015*; World Soil Resources Reports No. 106; FAO: Rome, Italy, 2014.

34. Seidler-Łożykowska, K.; Bocianowski, J. Evaluation of variability of morphological traits of selected caraway (*Carum carvi* L.) genotypes. *Ind. Crops Prod.* **2012**, *35*, 140–145. [CrossRef]

35. Dzieżyc, J. *Czynniki Plonotwórcze Plonowanie Roślin*; PWN: Warsaw, Poland, 1993; pp. 1–474.

36. Kaczorowska, Z. Rainfall in Poland over a long-term cross-section. *IG PAS Geogr. Work.* **1962**, *3*, 117.

37. Ziernicka, A. Classification of abnormalities in air temperature in southeast Poland. *Zesz. Nauk. Akad. Rol. Kraków., Rol.* **2001**, *22*, 5–18.

38. Kirchmann, H.; Bergström, L.; Kätterer, T.; Andrén, O.; Andersson, R. Can Organic Crop Production Feed the World? In *Organic Crop Production—Ambitions and Limitations*; Springer: Dordrecht, The Netherlands, 2008; pp. 39–72.

39. Murawska, B.; Piekut, A.; Jachymska, J.; Mitura, K.; Lipińska, K.J. Organic and Conventional Agriculture and the Size and Quality of Crops of Chosen Cultivated Plants. *Pol. Akad. Nauk* **2015**, *III/1*, 663–675. [CrossRef]

40. Sadowski, T.; Rychcik, B. Yield and chosen quality traits of oats grown in the period of conversion to organic cropping system. *Acta Sci. Pol. Agric.* **2009**, *8*, 47–55.

41. Wesołowski, M.; Cierpiała, R. The effect of the ploughed-in type of stubble catch crop on yield and weed infestation of organically grown oats. *Fragm. Agron.* **2013**, *30*, 133–144.

42. Klima, K.; Łabza, T.; Lepiarczyk, A. Participation of elements of cropping in the forming of the crop of glumiferous oats grown using traditional and organic systems. *J. Res. App. Agric. Eng.* **2014**, *59*, 115–118.
43. Klima, K.; Smaczny, M. Yielding and competiveness of oats and spring vetch depending on cultivation system and sowing method. *J. Res. App. Agric. Eng.* **2015**, *60*, 146–149.
44. Schrama, M.; de Haan, J.J.; Kroonen, M.; Verstegen, H.; Van der Putten, W.H. Crop Yield Gap and Stability in Organic and Conventional Farming Systems. *Agric. Ecosyst. Environ.* **2018**, *256*, 123–130. [CrossRef]
45. Raseduzzaman, M.; Jensen, E.S. Does intercropping enhance yield stability in arable crop production? A meta-analysis. *Eur. J. Agron.* **2017**, *91*, 25–33. [CrossRef]
46. Xue, Y.; Xia, H.; Christie, P.; Zhang, Z.; Li, L.; Tang, C. Crop Acquisition of Phosphorus, Iron and Zinc from Soil in Cereal/Legume Intercropping Systems: A Critical Review. *Ann. Bot.* **2016**, *117*, 363–377. [CrossRef]
47. Albayrak, S.; Güler, M.; Töngel, M.Ö. Effects of seed rates on forage production and hay quality of vetch-triticale mixtures. *Asian J. Plant Sci.* **2004**, *3*, 752–756. [CrossRef]
48. Lauk, R.; Lauk, E. Dual intercropping of common vetch and wheat or oats, effects on yields and interspecific competition. *Agron. Res.* **2009**, *7*, 21–32.
49. Pużyńska, K.; Pużyński, S.; Synowiec, A.; Bocianowski, J.; Lepiarczyk, A. Grain Yield and Total Protein Content of Organically Grown Oats–Vetch Mixtures Depending on Soil Type and Oats' Cultivar. *Agriculture* **2021**, *11*, 79. [CrossRef]
50. Polishchuk, V.; Zuravel, S.; Kravchuk, M.; Klymenko, T. Organic technology of growing vetch and oatsmixture under condition of using organic and mineral preparations under different fertilization systems. *Sci. Eur.* **2020**, *47*, 4–12.
51. Alemu, B.; Melaku, S.; Prasad, N.K. Effects of varying seed proportions and harvesting stages on biological compatibility and forage yield of oats (*Avena sativa* L.) and vetch (*Vicia villosa* R.) mixtures. *Livest. Res. Rural. Dev.* **2007**, *19*, 12.
52. Pisulewska, E.; Klima, K. Plonowanie wyki siewnej uprawianej w warunkach górskich w zależności od jej udziału w mieszankach z owsem. *Acta Agric. Silv.* **1999**, *37*, 77–85.
53. Behera, S.K.; Srivastava, P.; Pathre, U.V.; Tuli, R. An Indirect Method of Estimating Leaf Area Index in Jatropha Curcas L. Using LAI-2000 Plant Canopy Analyzer. *Agric. Meteorol.* **2010**, *150*, 307–311. [CrossRef]
54. Wu, S.; Huang, J.; Liu, X.; Fan, J.; Ma, G.; Zou, J. Assimilating MODIS-LAI into Crop Growth Model with EnKF to Predict Regional Crop Yield. In *Computer and Computing Technologies in Agriculture V, Proceedings of the International Conference on Computer and Computing Technologies in Agriculture (CCTA 2011), Beijing, China, 29–31 October 2011*; Springer: Berlin/Heidelberg, Germany, 2012; pp. 410–418.
55. Kwiatkowski, C.A.; Harasim, E. Chemical properties of soil in four-field crop rotations under organic and conventional farming systems. *Agronomy* **2020**, *10*, 1045. [CrossRef]
56. Zhao, B.; Ma, B.L.; Hu, Y.; Liu, J. Source–Sink Adjustment: A Mechanistic Understanding of the Timing and Severity of Drought Stress on Photosynthesis and Grain Yields of Two Contrasting Oats (*Avena sativa* L.) Genotypes. *J. Plant Growth Regul.* **2020**. [CrossRef]
57. Reiss, E.R.; Drinkwater, L.E. Cultivar Mixtures: A Meta-analysis of the Effect of Intraspecific Diversity on Crop Yield. *Ecol. Appl.* **2018**, *28*, 62–77. [CrossRef]